Shipwrecks and Global 'Worming'

P. Palma
L.N. Santhakumaran

Archaeopress

Archaeopress
Gordon House
276 Banbury Road
Oxford OX2 7ED

www.archaeopress.com

ISBN 978 1 78491 315 1
ISBN 978 1 78491 316 8 (epdf)

© Archaeopress, P Palma and L N Santhakumaran 2014

All rights reserved. No part of this book may be reproduced, stored in retrieval system, or transmitted, in any form or by any means, electronic, mechanical, photocopying or otherwise, without the prior written permission of the copyright owners.

Printed and bound in Great Britain by Marston Book Services Ltd, Oxfordshire

Contents

Abstract 1

Chapter 1. Introduction 3

Chapter 2. Historical Evidence 5

Chapter 3. Marine Wood-boring Organisms and their taxonomy 13
    Molluscan wood-borers: 14
    Shipworms (Teredinidae) 15
    Piddocks (Pholadidae: Martesiinae) 22
    Piddocks (Pholadidae: Xylophagainae) 24
    Crustacean attack 26
    Pill-bugs (Sphaeromatidae: Sphaeromatinae) 26
    Sphaeromatids 26
    Gribble (Limnoriidae) 27
    Chelura (Cheluridae) 29

Chapter 4. Wood-borer distribution 30

Chapter 5. Recent Findings 34
    Environment 36
    The Swash Channel Wreck 37

Chapter 6. Conclusions 40

Appendix I 44
    Systematic Treatment of Marine Wood-Borers 44
    Characters of Taxonomic Value for Identification of Marine Wood-Borers .. 47

Appendix II 52
    Check-List of Marine Wood-Borers 52

References 56

Paola Palma

Senior Lecturer, MSc Maritime Archaeology, Bournemouth University, Faculty of Science & Technology, Talbot Campus, Poole, Dorset, UK BH12 5B

L.N. Santhakumaran

Scientist-G (Retired), Institute of Wood Science & Technology, Bangalore-560003, India. (Present address : Aradhana, Patturaickal West, Thrissur-680022, Kerala, India).

Abstract

Marine borers, particularly the shipworms, as destroyers of timber, par excellence, are well known from very ancient times. They attacked the wooden hulls of ships with such intensity that the weakened bottom planks broke up even due to a mild impact caused by hitting a rock or any floating objects inducing shipwrecks. Even the survival of sunken ships as wrecks depends on the mercy of wood-destroying organisms, which may turn these 'port-holes' to history into meaningless junks. The silent saboteurs, involved in several early shipwrecks, are the molluscan and crustacean borers, aided by bacteria and fungi.

The paper presents an account of the marine wood-borers, together with a historical review of literature on their depredation on wooden ships, and on protective methods adopted from antiquity to modern times. The seriousness with which early mariners faced the problem of bio-deterioration and the fear the wood-borers created in their minds have been brought to light with, in some cases, excerpts from their journals and books. The anxiety and concern for protecting the ships from the ravages of wood-borers and for their own safety, as evidenced from their accounts, are discussed. Classification of various groups of marine wood-borers with notes on characters of systematic value and a complete list of species so far recorded in literature have been included under Appendix I and II. Methods employed to prevent damage to the boats included deep-charring, coating with pitch, coal-tar, whale oil and mustard oil with lime; scupper nailing ('filling'); sheathing with animal skin, hair, tarred paper, wooden boards (untreated or soaked in coal tar, Ferrous sulphate, Copper sulphate or Lead monoxide); sheathing with metals (Lead or Copper sheets); plastic, neoprene coated ply-woods; and painting with Copper oxide, Pentachlorophenol or phenylarsenious oxide. None of these imparts complete protection. Recent archaeological investigations carried out in British waters, especially on 'Mary Rose', are also summarised. It is suggested that, though borers are instrumental in inducing ship-wrecks thereby enriching the materials for archaeological studies, excavations at known ship-wreck sites should be augmented to unearth valuable historical data, before they are lost to satisfy the insatiable appetite of these pests.

Chapter 1. Introduction

Since antiquity, wood has been a material available almost anywhere and easily accessible to humans for a variety of different activities, from primary ones, such as hunting and cooking to secondary ones, such as exploration, sailing and trade. Timber, being the only constructional material available in the form in which it can be readily put to use, is naturally the first material employed by mankind for their varied activities, particularly for exploration of food resources in the sea, international trade and also for waging wars and piracy. Consequently, wooden craft, boats and ships have played a fundamental role in human ventures. However, the wooden hulls of these craft are prone to rapid infestation by certain types of marine organisms, such as wood-borers (shipworms, pholadids, gribbles and pill-bugs) and micro-organisms (bacteria and fungi), which attack the organic components of wood, acting as the 'primary active biological decomposers' (Brown, Bump and Muncher, 1988; Santhakumaran 1988). Thus, mankind had to confront the problem of marine bio-deterioration of timber, perhaps, from the very first day he set out into the sea on a wooden log or on a primitive craft. Of the above, the wood-boring molluscs of the *Teredinidae* family, commonly known as shipworms, are capable of rapid, high level degradation of wooden objects and their destructive potential to wood, especially to archaeological wood, is often underestimated, as the results of an attack can sometimes be deceptive, leaving wood looking externally sound, though internally timbers may be thoroughly honeycombed with tunnels (Santhakumaran 1988). (Figure 1)

FIGURE 1: BORER-INFESTED ARCHAEOLOGICAL WOOD FRAGMENT, SHOWING EXTERNAL SOUND APPEARANCE (LEFT) AGAINST THE INTERNAL HONEY-COMBED STRUCTURE (RIGHT).

Known to them as 'shipworms' or 'broma', all the ancient navigators had invariably a taste of the ruthless destruction caused by these borers to the hulls of their boats. The mariners were well aware of this hidden danger lurking in the underwater portions of their boats, and from their writings it was evident that they shuddered at the very thought of these organisms. Recorded history of early navigation has unfolded instances of several unhappy encounters between the navigators and marine wood-borers. Endowed with unlimited appetite ingesting any type of timber with apparent enthusiasm, prodigious fecundity resulting in heavy intensity of attack, fast rate of growth enabling the destruction of timber with remarkable rapidity, and highly specialized adaptations for boring into wood and leading a sedentary life, marine wood-borers are, from time immemorial, destroyers of timber par excellence and man's formidable enemy in the sea as agents of ancient ship-wrecks.

In the present paper, a historical review of literature on depredation of marine borers on wooden ships and on protective methods used from antiquity to modern times is presented together with an account of the organisms responsible for the damage. The role played by the marine wood-borers in causing ancient ship-wrecks, the fear these organisms created in the minds of early sailors and the seriousness with which they faced the problem of bio-deterioration have been brought to light, with, in some cases, excerpts from their Journals and books. Recent findings on archaeological wooden materials in British waters are also summarised.

Chapter 2. Historical Evidence

Normally a shipwreck can occur due to several reasons, for example as a result of an occasional fire while out in the seas, or due to rough and stormy weather conditions, when ships, out of control and struck on rocks or reefs, are swallowed by the very swollen waves, or sinking by enemies/pirates in battles or sabotage. But the recorded history of early voyages is replete with accounts of wood-borer infestation on ships, riddling the planks to the extent of causing shipwrecks. The consequences of their depredations on the hulls become unfortunately apparent only when the inevitable happens – breaking up of the weakened planks or springing up of catastrophic leaks – resulting in the ultimate disappearance of the ships even in moderately calm seas. Thus, the problem of bio-deterioration of timber in marine conditions is as old as the history of navigation.

One of the most known shipworms is *Teredo navalis* Linnaeus, with a wide spread distribution, not only in European waters, but also in the eastern and western coast of North America (Edmondson 1962).

The word Τερηδών is first to be found in the Greek playwright Aristophanes (*c*. 446 BC – *c*. 386 BC), but if there is some doubt whether was he specifically referring to the mollusc or a beetle – there is no doubt that the naturalist and philosopher Theophrastus (c. 371–c. 287 BC) with his extensive biology knowledge used this term, while referring to the shipworm and its degrading effects on wood in the sea (Jeffreys 1865). Also, in the comedy 'Knights' played in Athens in 424 BC, there is the mention of the wood-borer in a joking way 'I rather become an old maid here and be eaten by ship-worm' (Morrison et al. 2000), which implies that shipworms were a common knowledge as threat to wood vessels.

And if the organisms themselves were not such a common knowledge, the damage caused by shipworms to the hulls of ships (Santhakumaran 1990, 1991) was certainly a well-known problem to all ancient mariners since antiquity. A thorough compilation of most interesting evidences for these from original sources, (ship logs, travel logs, journals and books) is provided by Santhakumaran (1988, 1991). Amongst these, is the evidence from the following verse of the poet Publius Ovidius Naso (43 BC – 17 AD): '*Estur ut occulta vitiate teredine navis*' ('For as the ship by hidden shipworms spoiled') (Cnippingius 1670); and Pliny the Elder (23 – 79 AD) wondered and wrote 'What teeth, too, has she inserted in the teredo, to adapt it for piercing oak even with a sound which fully attests their destructive power! While at the same time she has made wood its principal nutriment'. The followers of Alexander the Great, have left a statement that, 'at Tylos, an island in the Red sea, there are trees, of which ships are built, the wood of which has been found uninjured at the end of two hundred years, even if it has been underwater all the time…'(Volume 3 of 1855 edition by H.T. Riley).

In his letter to the King and Queen of Spain, written from Jamaica on 7th July 1503, Columbus describes the havoc caused to his ships by marine borers (Colon 1503). 'On the last day of April, 1503, we left Veragua with three ships, intending to make our passage homeward to Spain, but as the ships were all pierced and eaten by the teredo, we could not keep them above water; we abandoned one of them after we had proceeded thirty leagues, the two which remained were even in a worse condition than that, so that all the hands were not sufficient with the use of pumps and kettles and pans to draw off water that came through the hole made by the worms. In this state with the utmost toil and danger we sailed for thirty-five days, thinking to reach Spain, and at the end of this time we arrived at the lowest point of the island of Cuba…' (Major 1847) (Veragua, a Province of Panama). Bishop (1913) states that 'of the four caravels of Christopher Columbus on his fourth voyage (1503), the *'Gallego'* and *'Biscaena'* were left behind at *Puerto Bello* (Portobello, Panama, in the Carribean), the remaining two, *'Capitana'* and *'Santiago de Palos'*, bored through and through by the Teredo, proceeded to Jamaica'. Even these two boats could hardly float because of the severe damage to their hulls and they also had to be finally abandoned. In fact, borers were so popular in the early sixteenth century that they were ignorantly connected with the evolution of geese. It was believed that, when wood is exposed in the sea, 'many worms generate in it and finally develop into geese with wings and feathers' (Boece 1527). Oviedo Y. Valdes (1547) states that 'Shipworms (broma) generate in wood…and honeycomb the planks so that they become like a sponge and do not hold water'.

According to the work of Roger of Hovendon, translated to English by Riley (1853), the ships, belonging to King Richard I of England, were badly mauled by marine wood-borers during the third Crusade, when he met King Philippe II of France at Messina, Sicily in 1190. 'The King of England in the meantime, while he was staying at Messina caused all the ships of his fleet to be hauled ashore and repaired, as many of them have become damaged in consequence of being eaten away by worms. For in the river Del Faro there are certain thin worms, which in the language of the people are called 'Boem', whose food is every kind of wood. Whenever these have once adhered to any kind of wood, they never leave go thereof, except through main force, until they have pierced right through; they make narrow straight holes when they have effected an entrance, and then from gnawing away the wood becomes so increased in size and bulk, that in coming forth they make wider holes' (Riley 1853, pp. 173-174).

Another instance of considering marine borers a real danger to early navigators is available in Adam (1599). While narrating an early voyage to Russia, it is stated that '...for the merchants, they get very strong and well-seasoned planks for the building, the Shippewrights, they with daily trauaile, and their greatest skill doe fitte them for the dispatch of the shippe: They calke them, pitch them and among the rest, they make one most stanch and firme, by an excellent and ingenious inuention. For thay had heard that in certain parts of the ocean, a kinde of wormes is bredde, which many times pearceth and eatheth through the strongest oake that is: and therefore that the mariners, and the rest to be imployed in this voyage might be free and safe from this danger, they couer a piece of the keele of the shippe with thinne sheetes of leade...'(Page 270 – 271 of the 1809 edition).

Also, Sir Richard Hawkins, while discussing the destructiveness of the 'arters' (shipworms) on his ship during the voyage to the South Sea in 1593, gives a very detailed description of a 'certain worm called broma' and its activity: 'for they enter in no bigger than a small Spanish needle, and by little and little their holes become ordinarily greater than a mans finger. The thicker the planke is, the greater he growth; yea, I have seene many ships so eaten, that the most of their plankes under water have beene like honey combes, and especially those betwixt wind and water. If they had not beene sheathed, it would have bin impossible that they could have swomme.'(Drinkwater Bethune 1847).

According to Moffett (1634), Francis Drakes ship 'Golden Hind', when returned to London in 1581, after the voyage around the world, had been rotten and spongy due to attack by *teredo*.

In fact, the maximum number of shipwrecks in one single incident, with marine wood-borers as the main culprits occurred during the debacle of the Spanish Armada in 1588 at the hands of the British Navy. Although the credit for this devastating blow to the Armada was shared by the British Navy and rough weather, the victors were unaware of their unseen allies in the form of marine borers indulged in their activities hidden in the hull planks of the Spanish Armada during its three month journey to the English Channel (Bitz 1967). By the time this impressive convoy of warships reached the battlefield, the borers they were carrying from Spanish waters, had done an excellent job of converting most of them into wrecks. Their boring activity was further aided by the warm water current in the Channel. After this defeat, the naval power of Spain was so much crippled that the King had to order his merchant ships to winter in the West Indies for fear of continued assaults from the British Navy. In 1590, Sir John Hawkins and Sir Martin Forbischer, in fact, undertook a voyage with ten British ships across the Azores to intercept the trade of Spain with the West Indies so as to further cripple their economy. In the late summer of 1591, when the Spanish ships attempted to return, nearly a hundred of them have already been converted into wrecks along with their rich cargo on account of intense borer infestation and damage, rendered easier when the ships were stationary, and sank (Monson 1682).

In the journal of Robert Boners, Master of the 'Dragon', about his voyage to the East Indies in 1611–1614, it is mentioned that 'I doe thinke that the Gulfe of Cambaya is the worst place in all the Indies for wormes, and therefore the ships which goe for Surat must have good provision' (Purchas 1625). (Gulf of Canbaya is Gulf of Cambay). Similarly, in Captain Walter Payton's journal on his second voyage to the East Indies in 1614, it is remarked that 'The double sheathing of ships which goe to Surat is of great purpose: for though the outermost sheathing be eaten like a honey combe with worms, yet the inner is not perished. It were also requisite that the Rudders were sheathed from thinne copper, to prevent the wormes eating off the edges thereof …' (Purchas 1625).

In the description of William Dampiers voyage around the world (1679 – 1691), one can see yet another example of how the borers had terrorized the early navigators. On his experience at Mindanao (S. Philippine islands) he writes 'About the middle of November, we began to work on our ships bottom, which we found very much eaten with the worm for this is a horrid place for worms. We did not know this till after we had been in the River a month, and then we found our Canoas bottoms eaten like honey combs; our Bark, which was a simple bottom, was eaten through, so that she could not swim. But our ship was sheathed, and the worm came no farther than the Hair between the sheathing plank and the main plank… We were told that in this place where we now lay, a Dutch ship was eaten up in 2 months' time… We had no worms until we came to this place… The Mindanayans were so sensible of these destructive insects, that whenever they come from sea, they immediately hale their ship into a dry-dock, and burn her bottom and there let her lye dry till they are ready to go to sea again'.(Dampiers 1697; Dampiers 1698: 362–363).

Godofredus Sellius, a Dutch writer is the author in 1733 of the *Historia Naturalis Teredinis Seu Xylophagi Marini, Tubulo-Conchoidis Speciatim Belgici : cum Tabulis ad Vivum Coloratis.* This is, in fact, the first edition of this important study on shipworms, several editions of which were published in various languages during the 18th century, as teredines were again considered a major threat to mercantile societies, like that of the Low Countries.

In his 'Essay on the marine of the ancients, particularly on their warships', Deslandes (1768) has mentioned that the ship belonging to Francois Cauche during his voyage from France to Madagascar (Indian Ocean), was attacked by worms and was made unseaworthy. Count D'Estrees lost six of his nine ships on a voyage from Brest (W. France) to Curacao (West Indies) and the remaining three returned with heavy worm attack. Similar accounts on problems of biodeterioration are available in Captain James Cook's journal of his voyage in the 'Endeavour'. A literal transcription of the original manuscript was brought out by Wharton (1893). The following citations will show how the borers with their depredations got the Captain and his crew preoccupied. 'May 24, 1769, at Otaheite. Having found the long boat leakey for these few days past, we hauld her ashore today to stop the leakes, when, to our great surprise, we found her bottom so much eaten by the Worms that it was necessary to give her a new one, and all the carpenters were immediately set to work upon her' (1893: 74). 'June 25, 1770, Queensland, Australia: Having run aground on a coral reef, the bottom of the 'Endeavour' was so damaged that some of the sheathing was stripped off. This alone will be sufficient to let the worm into her bottom, which may prove of bad consequence' (1893: 281). November 9, 1770, at Batavia: '...and found her bottom to be in a far worse condition than we expected, the false kiel was gone to within 20 feet of the sternpost, the main kiel wounded in many places very considerably, a great quantity of sheathing off, and several planks much damaged, especially under the main channel near the kiel, where 2 planks and ½ near 6 feet in length, were within 1/8 of an inch of being cutt through; and here the worms had made their way quite into the timbers, so that it was a matter of surprise to everyone who saw her bottom how we had kept her above water, and yet in this condition we had sailed some hundreds of Leagues, in as dangerous a Navigation as in any part of the world, happy in being ignorant of the continual danger we were in' (1893: 359). (The sheathing given to 'Endeavour' is probably of wood).

Narrating his experiences in the examination of ships in the Dockyard of Plymouth, England, Willcox (1827) reports that His Majesty's Ship 'Scepter, after leaving Bombay, India, for England in 1807, was obliged to return on account of a serious leak caused by *teredo* attacking a place where the copper sheathing had been damaged'. This ship was constructed of teak. The leak would have proved disastrous had it been developed in mid-ocean. Seymour remarks that 'no Russian ship in the Black Sea lasts more than ten years not only on account of the bad wood of which it is built, but also because of the worm *(Teredo navalis)* which infests Sevastopol and the southern coast of Crimea and commits great ravages among the ships' (1855: 92).

To add a final well-known example, the *Mary Rose* in England was no exception to this deterioration threat. Raised in 1982, the remains of the ship already showed damage by wood-borer action, where timbers had been exposed. In the Seventies, Alexander McKee and Dr Margaret Rule detected the presence of shipworms in the area where the *Mary Rose* sunk and after some research in collaboration with the Central Electricity Research Laboratories in Leatherhead, it was thought that the *Mary Rose* 'may have sunk because her timbers were rotten by *Teredo*' (McKee, 1973 p.245). In 2005, during some analytical work, a visual assessment of the deck beams on the *Mary Rose* confirmed a severe state of deterioration. Yet, it was not possible to conclude, if the attack had happened soon after the sinking or at any other time before the wreck had been discovered. But it could be hypothesised that the deterioration was not related to the most recent operations (Palma 2005a, 2005b).

In the five years from 1864 – 1869, about ten thousand sailing ships insured in England were lost in various parts of the world, nearly a thousand of them without any trace. The Portuguese lost 130ships between 1555 to1650on the route to the West Indies (Throckmorton 1970). Judging by the magnitude of the destructive power of these hidden enemies - the marine wood-borers - the reason for the disappearance of these early wooden ships is not difficult to deduce. The borer-ravaged bottom planks could easily crumble even due to a mild impact caused by hitting a rock or any floating object. Here is a group of organisms dreaded by early navigators and the very fact that they are known by different common names (pile-worms, shipworms or auger-worms to the English; tarets to the French; zeeworms to the Dutch; bromas to the Spanish; obe to the Fijians; cobra and warragara to the Australian aborigines; goon, kadi, tav, udhai, kalu, surali, yeel, kari, kali, lochok, kurli or thurli to fishermen along the west coast of India) itself demonstrates their 'popularity' among maritime people throughout the world. From the fore-going accounts, it is clear that marine wood-borers played the role of silent saboteurs of the sea, being instrumental in bringing about early shipwrecks. But the instances compiled above may not be even a fraction of the events that had happened in the history of navigation (Santhakumaran 1992).

A problem of vast scale, depicted in these examples, was and still is probably not just a physical and cultural problem, but was and still is an economical one as well. In the past, many have been the attempts put in place to try and contain the damage. Different approaches were taken to try and solve it – several records from written sources survive. Examples of anti-shipworm precautions taken are diverse, and across the centuries they included: wooden planks of the Kyrenia ship-wreck – a fourth century BC Greek merchant-ship – were found lead-sheathed to protect them from *Teredo* attack. Large sheets of lead were fixed and a wooden mallet, found in the ship, might have been possibly used for flattening the lead. Similar lead-sheathing with Copper tacks was also found on a fifth century BC ship-wreck in the Straits of Messina (Swiney and Katzeb 1973). The Spaniards and Italians also followed lead-sheathing as a protective method, in fact, ships belonging to Marcus Ulpius Nerva Trojanus (Roman Emperor 98–117 AD) were built of Pine and Cypress timbers covered with 'Greek pitch' and sheathed with lead plates fastened with copper nails (Deslandes 1768). According to Marco Polo, Chinese merchant ships were protected with a mixture of lime, chopped hemps and certain wood oil, which, then thoroughly amalgamated, held like a glue (Yule, 1903). The ship also had a double bottom with a sacrificial outer cover. Furthermore, written evidence describes that ships were taken into a dry dock and 'charring their bottoms to few millimetres (William Dampier's voyage around the world (1679-1691); Drinkwater Bethune 1847). Ship bottoms were also painted with tar or pitch or had their planks smeared with hot tallow, which went stiff and waxy, when chilled by seawater (Mountford 2002; Gianfrotta 2000).

In more tropical climate, a similar effect was achieved with lime and fish oil or lime and goat fat (Santhakumaran 1988). When possible, ships were run upriver into fresh water, where all marine growth and borers would die within a few days (Mountford 2002). Finally, Sir Hawkins British pride seemed to describe the most appropriate method to combat the shipworm problem: *'with thin bourds, halfe inche thicke; the thinner the better; and elme better than oake; for it ryveth not, it indureth better under water, and yeeldeth better to the shippes side...before the sheathing board is nailed on, upon the inner side of it they smere it over with tarre halfe a finger thicke and upon the tarre another halfe finger thicke of hayre, such as the whitelymers use, and so nayle it on, the nayles not above a spanne distance one from another; the thicker they are driven, the better. Some hold opinion that the tarre killeth the worme; other that the worme passing the sheathing, and seeking a way through, the hayre and the tarre so involve him that he is choked therewith; which me thinkes is most probably; this manner of sheathings was invented by my father, and experience hath thought it ot be the best and of least cost'* and 'covering the keel with a sacrificial timber called a 'worm shoe', (Mountford 2002); *'...with double planke, as thicke without as within, after the manner of furring which is little better than that with lead...'* (Drinkwater Bethune 1847); sheathing the hulls with lead or copper (Bingeman et al 2000) or fine canvas (Drinkwater Bethune 1847).

Among the many other remedial coatings employed in ancient times are: application of a mixture of lime and fish oil; goat fat and lime; deep-charring the outer planks and applying pitch; covering the hull with tar about 2 cm thick and spreading upon it a layer of animal hair and finally fixing thin planks (preferably elm) with nails; mixture of mustard oil and lime of shells; or pitch, brimstone and brick-dust or marble dust; or pitch, whale oil, tallow and glue; coating of pitch, tar or such bituminous compound with or without additives like powdered glass, cow's hair; covering the surface with tar, resin, soot, powdered charcoal, pyrites etc.; coating with tar and over it chenam (a mixture of lime and fish oil); mixture of pitch, tar and essence of tobacco; sheathing the hull with animal hyde, canvas, lead, copper or alloy of lead, antimony and quick-silver; driving broad-headed nails of copper or iron (called scupper-nailing or 'filling') nearly in contact with each other, so as to cover the entire hull after rusting; soaking the planks in solutions of ferrous sulphate, copper sulphate; application of oil and litharge (lead monoxide); washing the planks with a decoction of arsenic; and sheathing with fir, soaked or impregnated with coal-tar or paper dipped in tar. When double planking is employed, the space between the planks is filled with lime and fish oil, paper, cow hair, ashes, coconut fibres and cork. It is neither possible nor envisaged here to give a list of seemingly endless descriptions of combative techniques embarked upon by man against this animal foe in the sea (for further details on references in this regard, see Clapp and Kenk, 1963). Most of the above methods did not impart any protection at all and the few, which were slightly effective, also failed due to one reason or other, like poor adhesion to the hull, leaching action of the sea water, damage to the coatings when hit by a floating or a submerged object. Modern methods of protection of hulls of wooden boats include: use of copper naphthanate on the outside, followed by three coats of Copper bottom paint; sheathing with copper or aluminium or plastic casings; planking with resin-surfaced plywood or neoprene-coated plywood; and pressure-impregnation with wood preservatives, such as creosote, coal tar-fuel oil mixtures, Copper-Chrome-Arsenic (CCA) and copper-chrome-boron (CCB) (Kuenzel 1951; Santhakumaran et al.1984; Rao et al. 2007). A retention of 320 to 400 kg per cubic metre of oil type preservatives and 16 to 40 kg salts per cubic metre of water-borne preservatives is sufficient to impart prolonged protection from bio-deterioration in the sea (Santhakumaran et al.1984; Rao et al. 2007). The retention required depends on the end-use of the timber. For example, trials have shown that, for catamarans, where the craft is hauled up on the shore after day fishing, a load of 16 kg salts per cubic metre is enough to enhance the service life from a normal 3 to 5 years in the untreated condition to 10 to 15 years or even more. Though these methods are reportedly effective in preventing bio-deterioration, thereby prolonging the life of timber, when tried on simple craft like catamarans or on fixed piles, their application on ships and boats is bereft with practical problems with regard to their design and fabrication.

Chapter 3. Marine Wood-boring Organisms and their taxonomy

There are two groups of wood-borers in the sea. One group, closely related to oysters and clams (Mollusca: Bivalvia), is called 'shipworms' (Family: Teredinidae) and 'piddocks' (Family: Pholadidae) (Figure 2A,B,C). The other group, distant cousins of prawns and crabs (Arthropoda: Crustacea) is known as 'pill-bugs' (Family: Sphaeromatidae) and 'gribble' (Family: Limnoriidae) (Figure 2 D,E). (A third crustacean family, Cheluridae, also includes wood-dwelling members making long furrows on the wood surface. They are usually found in association with *Limnoria* in the littoral zone and are not very important from wood destruction point of view).

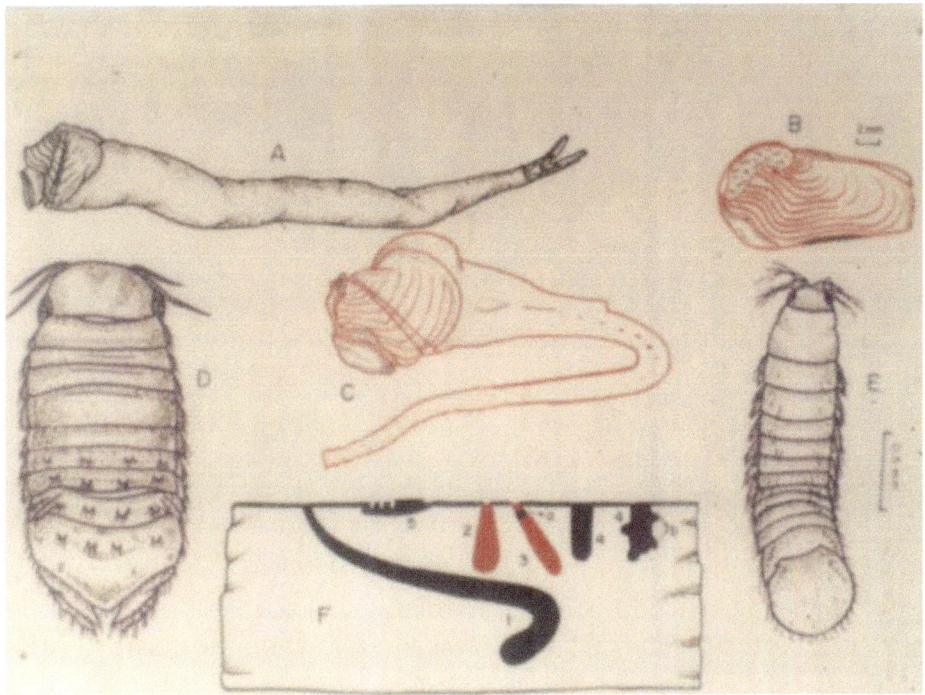

FIGURE 2: DIFFERENT TYPES OF MARINE WOOD-BORERS.
A:SHIPWORM; B:PIDDOCKS (*MARTESIA*); C:PIDDOCKS (*XYLOPHAGA*); D PILL-BUGS (*SPHAEROMA*); E GRIBBLES (*LIMNORIA*); AND F:BURROWS PRODUCED BY THEM (F1:SHIPWORM; F2: PIDDOCK (*MARTESIA*); F3 : PIDDOCK (*XYLOPHAGA*); F4:PILL-BUG; F4B :PILL-BUG WITH JUVENILE SIDE BURROWS; AND F5 :GRIBBLE

For the purpose of this article the shipworm category is the one mainly considered due to the high level of damage caused to the archaeological wood materials on the seabed exposed to aerobic conditions. Other groups are only briefly mentioned, as they also cause damage to the wood surface and in the light of recent discovery of the existence of several new species in very deep waters in numbers large enough to destroy wood debris. – therefore potentially destroying details such as axe marks and carved details.

Molluscan wood-borers:

FIGURE 3 PINE-WOOD PANELS, DESTROYED BY MARINE BORERS OM INDIA.
NOTE THE COUNTLESS TINY LARVAL ENTRY HOLES ON THE PANEL.

Molluscan wood-borers are permanently entombed within the substratum, as they enter the wood while in the larval stage. The only indication of the internal presence of large number of adults is the numerous small pin holes on the wood surface, rendering the assessment of the extent of destruction inside impossible (Figure 3). As mentioned earlier, molluscan borers collectively belong to the common group shipworms (Family Teredinidae) and piddocks (Family Pholadidae).

Shipworms (Teredinidae)

As shipworms are the cause of the most extensive wood degradation in the marine environment, it helps to understand their anatomy and, therefore, the way they cause such an extensive damage.

For a long time, classification of Teredinidae species (shipworms) was based entirely on shells and pallets, although variations in specimens of the same species were confusing. It was Turner (1966), who brought out the importance of the anatomy of shipworms in the systematics of this group, particularly in generic classification. Characters of systematics value for species identification are nature of the shell valves, tubes (internal lining of the borrows which sometimes gets thickened as a tube particularly at the posterior end), pallets (a pair of calcareous organ situated at the posterior end of the animal which is used to plug the entry hole during adverse conditions or when the borer is disturbed) and siphons. Of these, the morphological variations exhibited by the pallets are remarkable and almost all the species can be identified from their pallets (Turner 1966, 1971). Other characters are of limited help, but when considered with more important characters, may prove useful in separating closely related species. It will be useful to examine a series of pallets of the same species, as fresh as possible, and some characters are better discernible under transmitted light.

Shipworms have an elongated worm-like body (Figure 4) with the two shell valves and cephalic hood at the anterior end and with the siphons and pallets at the posterior end. The soft body with fundamental and vital organs are not enclosed by the shell, as in a typical bivalve. The remark of Sellius (1733) that the shells are too fragile and are not powerful enough to bore in the wood goes against the study of several subsequent authors who – with the occasional difference in interpretation - accepted that 'the foot of *Teredo* adheres to the wood acting as a centre-bit, while the animal is at work with the shell' (Home 1806; Miller 1924).

The entire body, though not enclosed by the shell valves, is well protected within the burrow. The interior of the tunnels is coated with a calcareous substance secreted by the organism itself (Calman 1919). The construction of this is started by the larvae at its phase of entering the wood, but at this stage, it looks like a minuscule mound with two pin head-size holes for the siphons to protrude on the wood surface. This has the function of protecting the larvae while digging the tunnel in the wood, yet leaving space for the few hour old siphons to wave in the water and perform normal living activities, such as breathing and waste release. In the second stage, as the shipworm grows in size, it enlarges the burrow and, thus, the borer is well protected inside the wood and its only connection to the exterior world is through the siphons waving in the water constituting a potential pray for fish swimming in the proximity eager to graze these appendages. But, when attacked by a predator, the borers are capable of withdrawing the siphons inside and plug the entry hole with a pair of calcareous organ called pallets.

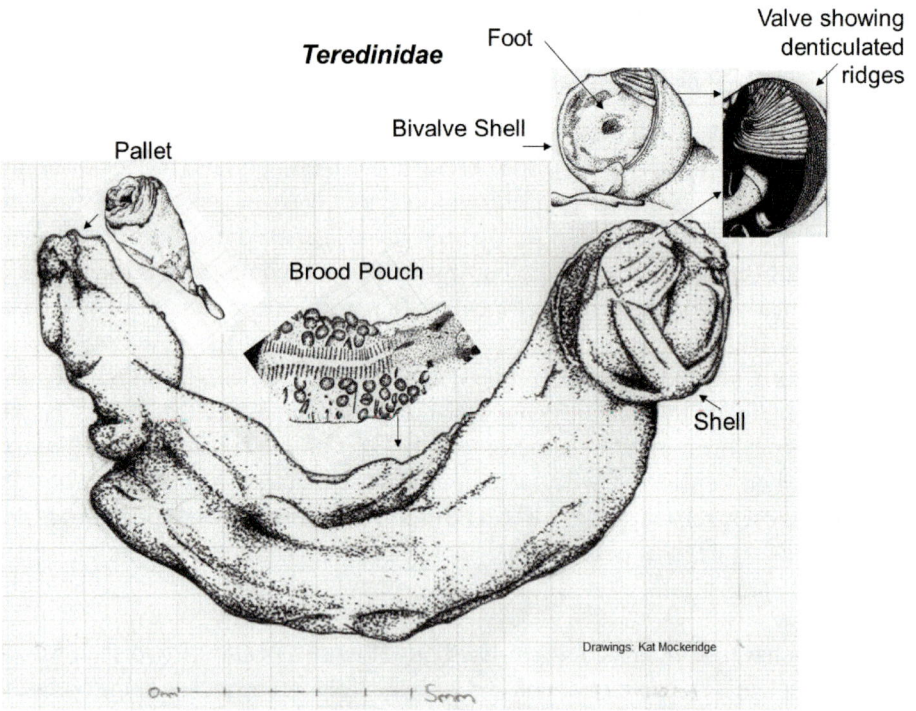

FIGURE 4: MORPHOLOGY OF A SHIPWORM. (DRAWING: KAT MOCKERIDGE)

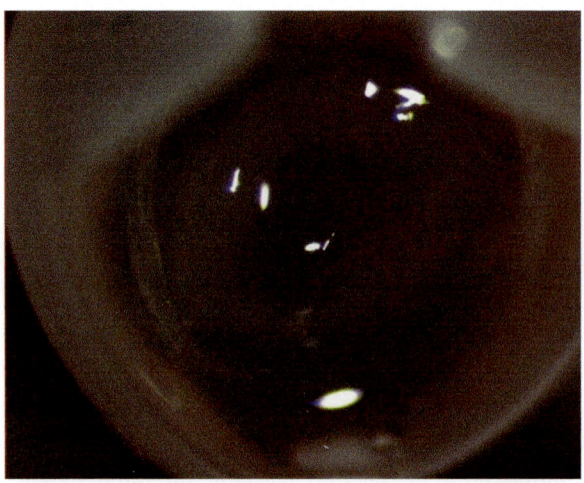

FIGURE 5: MUSCULAR FOOT IN BETWEEN THE SHELL VALVES.

The shell valves of the shipworm are responsible for rasping off wood and tunnelling into the wood, with a rotating and forward movement and are kept in place by the suction of the muscular foot (Figure 5), which is adhering to the blind end of the burrow. The two valves have denticulated ridges used in a grinding action (Figure 6). These ridges – which under the microscope look like sharp serrated knives - are also a record of the growth rate and, therefore, age of the animal, as well as detailing the conditions under which it lived (Miller 1922; Calman 1919), as their deposition is affected by food, salinity and temperature.

FIGURE 6 ELECTRON MICROGRAPH OF A PORTION OF THE SHELL VALVE MAGNIFIED TO SHOW THE TOOTHED RIDGES.

The action of micro flora, fungi and bacteria in softening the waterlogged wood surface and of other burrowing species, such as some small isopods (Figure 2 D,E) is exploited by shipworms to enter and remain in the wood, yet shipworms are also able to penetrate the surface independently without such pre-conditioning (Board & Feaver 1973; Isham & Tierney 1953). *Teredo* enters the wood at a larval stage of about 55 – 85microns in diameter and never leaves it. The metamorphosis of *T. navalis* from larva to adult takes about ten days (Hurley 1959). As the shipworm grows, it enlarges and increases the burrow length sufficient enough to accommodate the entire body. The lining of the burrow walls with a calcareous secretion of the mantle protects the borer from the harmful effects of any toxic chemicals present in the wood (Santhakumaran 2005). It is to be believed that '*The thicker the planke is, the greate the groweth*' (Drinkwater Bethune 1847; Board & Feaver 1973), indicating that the bigger the timber the more damage it would suffer.

Shipworms live in tunnels, which they dig, when they come into contact with the surface of wood. They remain in these tunnels for the duration of their lives because, once they have entered it, they are unable to leave it, as the organism grows inside the timber structure and its size is bigger than the initial entry hole, where pallets and siphons are located (Eldman Abbate 1961). The calcareous lining of the burrows forms a multi strata layer in some species or gets highly thickened depending on the chemicals present in the wood. Sometimes these calcareous tubes extend up to about 30 mm outside the wood substratum in specimens growing in polluted waters or in such adverse environmental conditions. They can be then easily mistaken for serpulid tubes (Santhakumaran 1978) (Figure 7).

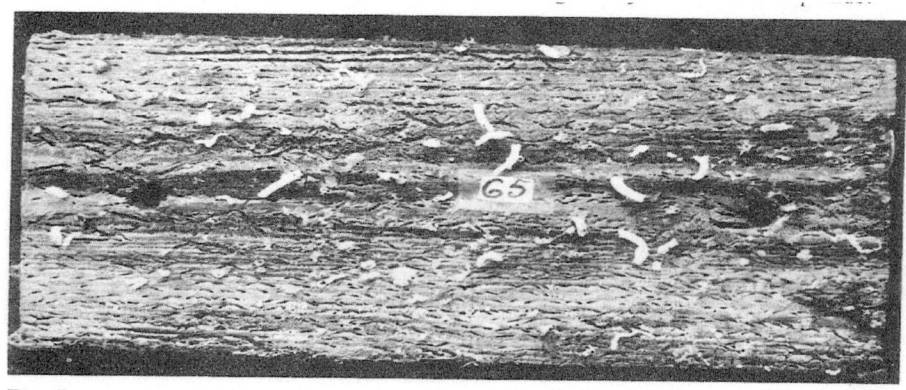

FIGURE 7 CALCAREOUS TUBES OF SHIPWORMS PROTRUDING FROM A PINE WOOD, LOOKING LIKE SERPULIDS. NOTE THE HEAVY INFESTATION OF *LIMNORIA*. (LOCALITY : TRONDHEIM, NORWAY).

The lining is not extended to the very blind end of the tunnel, as the shipworm needs its shell free to dig into the wood, which it achieves by making rapid rotary shell movements. It is this tunnelling action which, in 1818, was thought to have inspired Marc Isambard Brunel's design for a moveable shield for constructing tunnels; a method still in use today. Shipworms enter the wood perpendicularly, but once inside, soon follow parallel to the wood grain in a longitudinal orientation (Goodell 2000; Lopez-Anido et al. 2004). They do not bore through a neighbouring tunnel and, when the organisms detect the presence of other shipworms in its trajectory, they turn and change the direction. Thus, the tunnels take a zigzag course inside the wood (Figure 8).

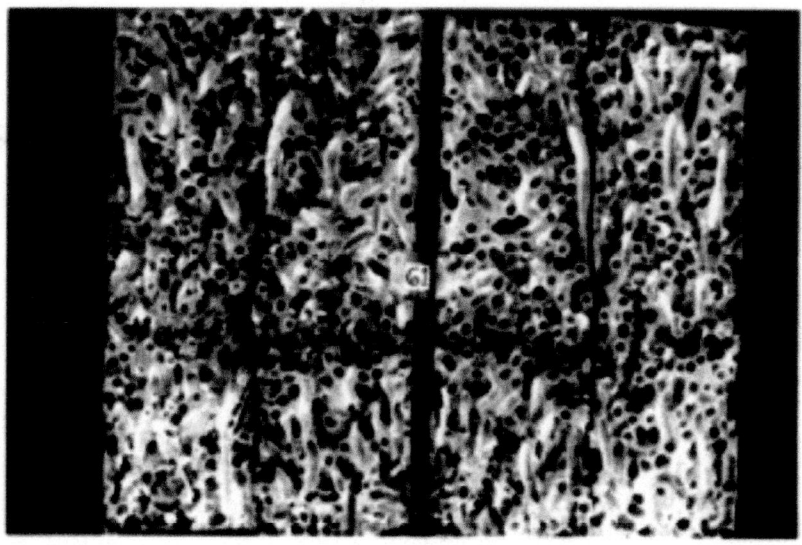

FIGURE 8 A PANEL OF PINUS SYLVESTRIS, SPLIT OPEN TO SHOW HEAVY INTERNAL DESTRUCTION BY SHIPWORMS WITHIN 12 MONTHS (LOCALITY : TRONDHEIM, NORWAY).

As mentioned earlier, shipworms are connected to the outside water by a pair of siphons (Figure 9), extended through the tiny larval entry holes, which enable the borers to perform two functions; firstly, to bring in sea water through the inhalant siphon for respiration and also to consume the planktonic organisms present in it to supplement their nutritional requirements, and secondly, for releasing waste materials along with the water discarded through the exhalant siphon.

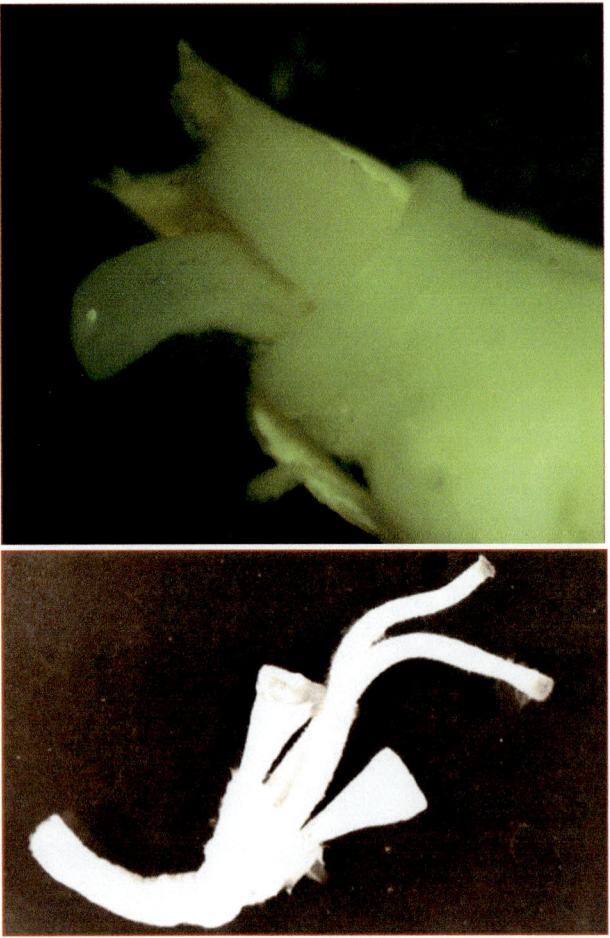

FIGURE 9: A) LATERAL VIEW OF THE POSTERIOR END OF A SHIPWORM, SHOWING ONE OF THE SIPHONS AND THE PAIR OF PALLETS. B) : POSTERIOR PART OF A SHIPWORM (*DICYATHIFER MANNI*) SHOWING EXTENDED SIPHONS AND THE TWO PALLETS.

If the organisms recognise a potential threat – changes in the temperature or when the salinity of the water is sufficiently low (Blum 1922) or a predator or presence of toxic pollutants, the siphons are retracted into the tunnel and the pallets – two small calcareous appendices - plug the hole, just like a door, closing off the outside world (Figure 9). The borer can remain alive for several days inside its tunnel, but if the salinity level falls less than 5 parts per 1000 for a long period, this constitutes a high risk for the specimens (Blum 1922).

The morphological variations, exhibited by the pallets, are remarkable and almost all the species can be identified from the nature of their pallets (Turner 1966, 1971)

(Figure10). Classification is based mainly upon their size, shape and colour (Calman 1919). They are fragile structures, easily damaged and degraded by age. In addition to variation of the genus, the appearance of the pallets may be greatly affected by ecological and environmental conditions (Miller 1923).

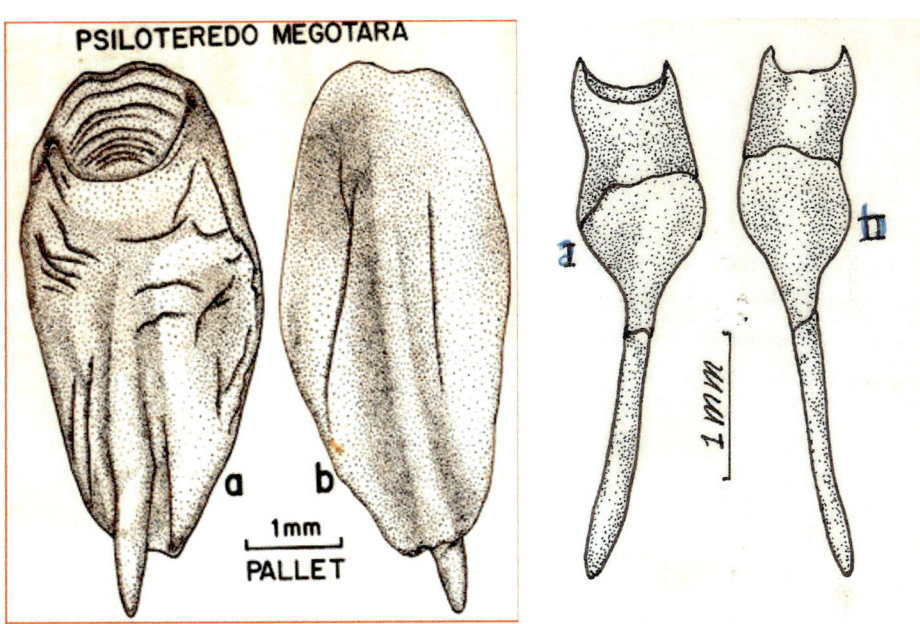

FIGURE 10: PALLETS OF TWO SPECIES (*PSILOTEREDO MEGATORA* ON THE LEFT AND *LYRODUS PEDICELLATUS* ON THE RIGHT) OF SHIPWORMS SHOWING VARIATION.
A : OUTER VIEW; B : INNER VIEW.

About 75 species of shipworms have so far been described from all over the world (Turner, 1966, 1971; MacIntosh 2012), the Teredininae and Bankiinae being sub families accommodating them (See Appendix – II for list of species). Important species, recorded from wooden hulls, are: *Lyrodus pedicellatus* (Quatrefages) and *T.navalis* (both belonging to Teredininae), *Nototeredo norvagica* (Spengler) and *Psiloteredo megotara* (Hanley) (both Bankiinae) in European waters; *L. pedicellatus* and *Teredothyra matocotana* (Bartsch) (Teredininae) from Pacific Ocean; and *Bankia campanellata* Moll & Roch, *Bankia rochi* Moll, *Bankia carinata* (Gray) (all Bankiinae), *Dicyathifer manni* (Wright), *L. pedicellatus*, *Teredo clapi* Bartsch, *Teredo furcifera* von Martens and *Nototeredo edax* (Hedley) (all Teredininae) from Indian waters. In fact, the number of species involved may be much more, as in several instances, the borers encountered in shipwrecks have been collectively recorded as 'shipworms' without, unfortunately, attempting species identification.

T. navalis is probably the shipworm species best-known to archaeologists. It's a highly specialized bivalve, a worm-like mollusc, adapted for boring into wood. This particular borer has a fast growing period reaching in certain circumstances a length of 100 mm in a month (Turner 1947) and, in certain regions can its length vary between 15 cm to 180cm, with a diameter up to 2.5 cm (Chellis 1961; Lopez-Anido et al. 2004), depending on the species, environmental characteristics, size and nature of the substratum and the conditions of food-supply, temperature and salinity, all affecting not just the size but the growth rate as well (Calman 1919).

Routine examination of the infested wood under the microscope will reveal the presence of numerous tiny entry holes of the shipworms. By this method, however, it is not possible to ascertain the extent of internal damage caused by the borers, for which splitting open the wood, thereby sacrificing or destroying the sample, is necessary. This is not often possible, if the wood has any archaeological importance and needs preservation. Therefore, if small holes are seen under the microscope confirming the infestation by shipworms, an x-ray investigation is essential to assess the internal condition, thus saving such specimens of archaeological value for preservation for posterity. X-ray examination is thus a non-destructive method to determine the general state of internal attack (Oliver 1961). X-ray photographs will reveal the calcareous parts of the borers, as well as the directions of calcareous lining of the tunnels inside the wood indicating the extent of attack and internal damage. Although it is possible to reveal different levels of degradation, an x-ray's usefulness in detecting early stages of attack is limited, especially while examining thicker wood pieces. As far as possible, detailed visual inspection under a microscope must be relied upon to obtain actual picture.

Both microscopy and x-ray analysis were utilised during the 3-year MoSS (Monitoring of Shipwreck Sites) European Research Project[1], where wooden panels were exposed in the sea column and retrieved at different intervals, to investigate how quickly they were attacked and by which organisms. This project set the standard for a procedure, which could be applied at different sites across the globe. The research was supported by the use of data loggers, deployed to retrieve information about the environmental parameters, such as temperature, dissolved oxygen, pH, Redox Potential and depth at each site (Palma 2005c).

Piddocks (Pholadidae: Martesiinae)

Wood-boring members of the pholadid sub-family, Martesiinae are typical bivalves, and unlike shipworms, the soft body is enclosed by the two shell valves (Figures 2B, 11,12).They are found in the coastal waters of tropical and subtropical seas. Boring into wood is accomplished with the toothed ridges on the shell valves, as in the case

---

[1] www.mossproject.com

of shipworms. Here also, fresh attack is initiated by the young larvae, and adults, grown inside the burrow, cannot come out of the wood. Therefore, internal damage cannot be assessed without splitting open the wood. Specimens grow to a size of 2 to 4 cm, producing superficial pear-shaped burrows almost perpendicular to the grain of the wood. Though intensity of attack increases with depth, variations to this pattern have also been reported. They cannot digest wood cellulose and subsists on plankton. Sexual maturity is attained at an early age (even within two months in tropical waters), producing large number of eggs in one brood (Santhakumaran 2005).

Eight species of Martesiinae are known to infest wood (Turner 1971; Turner & Santhakumaran 1989). (See Appendix- II for list of wood-boring species). Most widely distributed and cosmopolitan species in tropical seas is Martesia striata Linnaeus (Figure 11),which has been recorded from the hulls of boats and ships. Since the attack is only for protection, and the animal does not consume the sawdust produced during boring, no timber is naturally resistant to its attack and the species is quite destructive and is a threat to timber exposed in the littoral seas. Allied species, Martesia nairi (Turner & Santhakumaran) (Figure12) is generally found in large intensity only in mangrove habitats destroying trash wood and even live vegetation, and Martesia fragilis Verrill & Bush mostly attacks floating nuts and drift-woods and is of extremely rare occurrence in wooden substratum and, hence, insignificant from the wood destruction point of view).

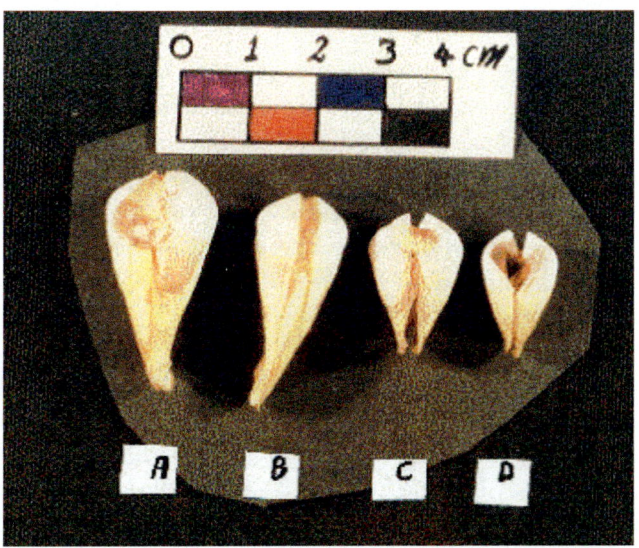

FIGURE 11: ENTIRE SPECIMENS OF *MARTESIA STRIATA*. A : ADULT, DORSAL VIEW; B : ADULT, VENTRAL VIEW; C : JUVENILE, DORSAL VIEW; AND D : JUVENILE, VENTRAL VIEW.

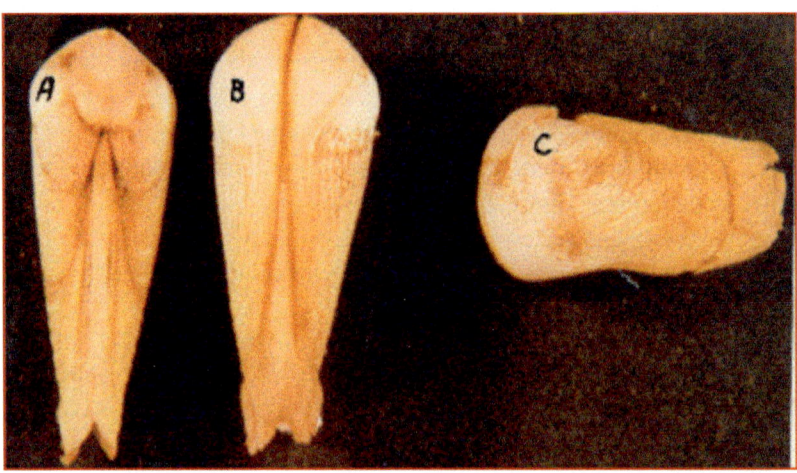

FIGURE 12: ENTIRE SPECIMENS OF *MARTESIA NAIRI*. A : DORSAl VIEW; B : VENTRAL VIEW; C : LATERAL VIEW

Piddocks(Pholadidae: Xylophagainae)

Members of the subfamily, Xylophagainae, are also typical bivalves (Figure 2C,13) with the soft body almost completely enclosed by the two shell valves (except in *Xylopholas* and *Xyloredo,* where the body is elongate and teredinid-like). As in the case of Martesiinae, here also pallets are absent, but the soft body is well protected within the pair of shell valves (except in *Xylopholas* and *Xyloredo,* which combine morphological features of a pholadid and teredinid). Posterior end of the burrow is filled with a tube, called chimney, formed of agglutinated particles of frasse or of particles broken off during boring but not ingested (Figure 2F^{3a}). The siphons protrude through these tubes. (In *Xylopholas,* the soft body outside the shell valves is protected by a periostracal sheath, which terminates in a pair of sub-oval, slightly calcified lateral plates; and in *Xyloredo,* the protection to the teredinid-like body outside the shell is afforded by a thin calcareous lining).

Burrows are long and pear-shaped without calcareous lining (except in the genus *Xyloredo*) on the walls. Attack is superficial, perpendicular or sometimes oblique to the grain of the timber (Figure 2F^3). Some species grow to a shell length of 23 mm (Burrow length may be 50 to 70 mm). In Xylopholas and Xyloredo, however, the burrows are long and deep inside the wood. Xylophagainids can utilize the wood as food. (Mechanism involved in cellulose digestion is not known, but is probably via bacteria and phagocytes. Like teredinids, a storage pouch (caecum) of the stomach is present).

Sexual maturity is attained at an early age (even within one month). An accessory genital organ, vesicular seminalis, is present for the storage of sperms, probably allowing self-fertilization (a probable adaptation to deep sea existence). Members are oviparous and release large number of eggs in one brood. (In some deep-water species development takes place within the burrow and young ones are retained up to late veliger stage. In several instances, larvae have been found attached to shell or siphons of the parents). Shallow water forms breed during summer and early autumn. Sexually mature individuals are present throughout the year. Intensity of attack increases with depth, with maximum near the mud-level. However, intensity of infestation will be much greater on test blocks installed near the mud levels at 5 m and 10m depths than on those suspended at 5 m and 10 m depths in a water column, where the depth is 15 m or more (Santhakumaran 1984). Dispersal of species is only through the agency of infested wood, as adults cannot come out of their tunnels. The adults release the eggs on arrival in a new locality, which, in turn, infest fresh timber structures. The developing eggs and larvae are also carried to some distance by waves and currents. (Some of the above characters are based on laboratory observations on *Xylophaga dorsalis* (Turton) (Figure 13) and *Xylophaga praestans* Smith and cannot be generalized for all species. Very little information is available on the biology of deep-water species).

FIGURE 13: *XYLOPHAGA DORSALIS*, PHOTOGRAPHED ALIVE.

Members of the sub-family Xylophagainae replace *Martesia* in higher latitudes and in deep sea. Few species occur from shallow waters to great depths and in some localities, xylophagainids are as destructive as teredinids and *Martesia*.

Almost all the species of xylophagainids have been collected and described from woody materials dredged out from great depths, though, interestingly, at least sixteen new species were recently added from deep-sea deployments installed at eastern and north-east Pacific Ocean (Harvey 1996; Knudsen 1961; Turner 1972a, 1972b, 2002; Voight 2007, 2009). Pailleret et al. (2007) have recorded *Neoxylophaga* teramachi Taki and Habe, along with the shipworms *L. pedicellatus* and *T. matocotana*, from samples of sunken wood (*Fitchia* sp., *Fuchsia* sp., *Leucaena* sp.and *Serianthes* sp.), collected from 560 to 580 metres depth, off the Vanuatu Islands.

About fifty-eight species of xylophagainids are so far reported as wood-borers from different parts of the world. (Harvey 1996; Knudsen 1961; Okutani 1975; Santhakumaran 1980, 1984; Taki and Habe 1950; Turner 1971, 1972a, 1972b, 2002; Voight 2007, 2009) (See Appendix – II for list of species).

Crustacean attack

Crustacean attack is usually more detectable than that of the shipworm, as the tunnels, excavated by these borers, are mainly superficial on the wood surface, creating an interconnected system of tunnels, where it lives in couples up to a period of 10 months or more. Borers belong to the families Sphaeromatidae (pill-bugs), Limnoriidae ('gribbles') and Cheluridae.

Pill-bugs (Sphaeromatidae: Sphaeromatinae)

Sphaeromatids (Figures 2D, 14) are generally well distributed in the estuaries and backwaters, particularly in the tropics, where they cause extensive superficial damage to timber structures in their intertidal portion (Figure 18). They are also found in wood lying in the littoral waters. Burrows are cylindrical, almost perpendicular to the grain of the wood and measures 2 cm to 4 cm in length (Figure 2F4). Borers attain a length of 8 mm to 15 mm. They produce 10 to 15 eggs (or rarely more) in one brood.

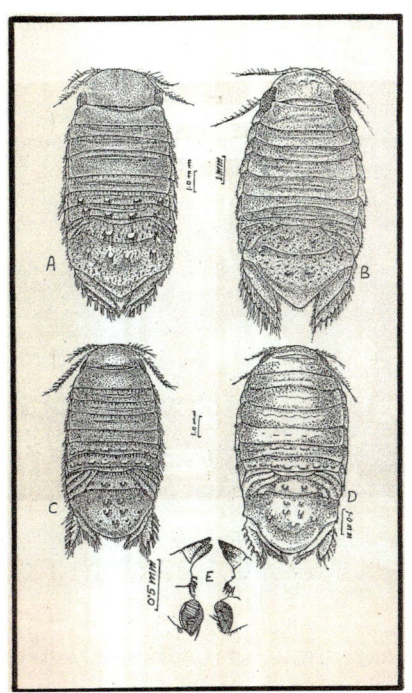

FIGURE 14: DIFFERENT SPECIES OF PILL-BUGS. A : *SPHAEROMA TEREBRANS*; B : *SPHAEROMA TRISTE*; C : *SPHAEROMA ANNANDALEI*; D : *S. ANNANDALEI TRAVANCORENSIS*; E :MANDIBLES.

After pairing, fertilized eggs are retained by the female in an internal brood pouch, where the eggs develop and young ones are released. Juveniles either come out and make a fresh burrow or start burrowing from the parent burrow itself. Dispersal of the species is through infested wood or through the active migration of individuals. They can tolerate salinity as low as $1°/_{oo}$ (0.1%) and even fresh water. In all, about 11 species of wood-boring pill-bugs are reported in literature (Harrison and Holdich 1984) (See Appendix – II for list of species). Species so far recorded from boat hulls are *Sphaeroma annandalei* Stebbing and *Sphaeroma terebrans* Bate. Because of their preference to estuarine and mangrove habitats, sphaeromatids may not be important in destroying submerged wood from ship-wrecks in greater depths.

Gribble (Limnoriidae)

The *Limnoria* galleries are very superficial, narrow and run parallel to the wood grain (Figure 2F5). The young ones that are released start tunnelling from the parent burrow itself and, therefore, *Limnoria* tunnels will have a branching pattern and, when the attack is heavy, the wood surface will have a lace-like appearance (Figure 15).

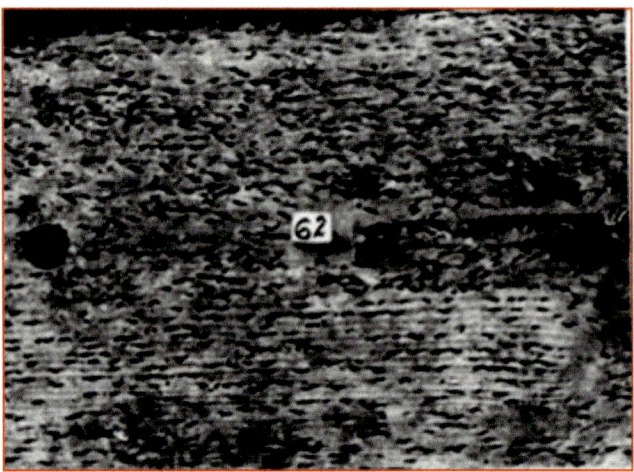

FIGURE 15: PINEWOOD PANEL SHOWING HEAVY INFESTATION BY LIMNORIA IN LESS THAN ONE YEAR. (LOCALITY : TRONDHEIM, NORWAY).

Body of *Limnoria* is grey or off-white, and segmented and its length varies between 2-8 mm depending on the species. Only a small number of eggs are laid, up to 30, depending on the species – a sharp contrast in the case of both isopod borer families when compared to the oviparous shipworms' breeding capabilities. (Shipworms release thousands of ova in each brood, obviously to give allowance to the large-scale mortality of the developing larvae during its planktonic life, while in the crustaceans, the young ones are well protected within the brood pouch of the adult. In the case of those genera of shipworms, which exhibit vivipary, the young ones produced are much less). It is distributed mainly through infested driftwood or wooden boats, as its ability to swim is very limited. Nevertheless, in this mode of transport, the borer can obviously reach considerable distance in a relatively short period of time. Though much smaller than members of the other isopod wood-borer sub-family Sphaeromatinae, limnoriids too, because of high rate of infestation, cause extensive damage to underwater timber structures, particularly in higher latitudes. Earlier they were considered not very important in tropical waters along Indian coasts. However, during the past two decades, they were found to have well spread along the east and west coasts of India with severe infestations in several localities, *Limnoria indica* Kampf & Becker and *Limnoria platycauda* Menzies being the predominant ones, posing a potential threat to the survival of hundreds of shipwrecks waiting to be excavated from the country's vast near-shore areas.

There are about 31 wood-boring limnoriid species (Cookson 1991; Cookson et al. 2012). *Limnoria lignorum* (Rathke) has been recorded from shipwrecks in European waters. When salvaged in 1982, wooden components of the war-ship, Mary Rose, which sank in 1545, indicated severe *Limnoria* attack, and borers were present even on some of the arrows in the armoury.

Chelura (Cheluridae)

The *Chelura* is pale pink to orange-red, measuring between 3-8 mm in length. The most common species is the *Chelura terebrans* Phillippi. It is generally a native of temperate waters and is widely recorded on the South coast of England and Italian coasts. It produces a small number of eggs (between 5-15 eggs). The damage, caused by the number of galleries (furrows) dug by these borers, represents a very small threat to wooden artefacts or structures. They are found in nature in association with *Limnoria,* often occupying and enlarging the latter's tunnels and spreading the destruction of wood; in fact, *Chelura* is known to feed on both *Limnoria* as well as its own faecal pallets (Kuhne & Becker 1964, Cragg et al. 1999). However, it has been reported that this borer possesses enzyme mechanism capable of digesting wood cellulose.

Four species of chelurids are reported in literature (Barnard 1959; Ortiz 1976; Shino 1957). (See Appendix – II).

The damage caused by the crustacean borers also is well studied, yet the control of crustacean borers too still remains an unresolved problem (Cragg et al. 1999) and a lasting method to combat their destructive action has still eluded the investigators.

Chapter 4. Wood-borer distribution

The presence of shipworms on a specific site is ruled by several different as well as coinciding characteristics and variables, such as temperature, salinity and, more importantly, the presence of wood. Other factors, like tides, currents, turbidity, pollution and nature and composition of associated biota, also contribute to their distribution and presence, but are of local effect (Turner 1966). Here again, different species of timber exhibit different degrees of resistance to marine wood-borers. A number of papers has been published since the early 1900s on the natural durability of wood, with a view to screening out species useful for the marine construction sector (Clapp & Kenk 1963; Edmondson 1955; Santhakumaran 1994; Santhakumaran & Alikunhi 1983). The fear marine wood-borers created in the minds of early sea-farers is also amply reflected in literature describing the latter's preferences for certain type of timber species on the false assumption that some of them are more resistant to bio-deterioration. References are available to show the extensive use of beech, oak, ash, mulberry, elm, fir, pine and cedar in western countries (Steinmayer Jr and Turfa 1996) and teak, benteak, sissoo, Indian Rose-wood, *Terminalia* spp. and *Phoebe* spp. in the tropics (Sila Tripati et al., 2005 and 2009). From the archaeological point of view, perhaps, the most widely used species in shipbuilding are *Quercus rubra* (oak), *Pinus sylvestris* (Scots Pine) *and Elminus modestus* (Elm). Yet, even for these species their durability depends on the location, the sediment characteristics, temperature and salinity and consequently also on the types of wood-borers present at the site. Such selective usage is presumably based more on the availability of timber species and on fragmented experience than to their durability to withstand marine borer attack, derived from experimental data. As previously mentioned, because of the complexities of the problem of marine bio-deterioration, no timber species can escape the destructive activities of marine wood-borers. Thus, the survival of the wood is very much dependent on all of the above variables.

It is worthwhile here to discuss the depth range of marine wood-borers to ascertain their distribution and the threat paused by them to wrecks lying in greater depths. Expectedly, the presence of wood materials and length of free-swimming larval stages decide the distribution of borers in littoral and deep waters. The vertical distribution of different types of borers is as follows: The shipworm infestation increases with increasing depths and maximum incidence and destruction of wood are noticed near the mud level. In the case of the tropical pholadid species, no definite pattern of depth preference is observed for settlement. Though, generally, they also prefer the deeper levels, their pattern of vertical distribution reportedly differs at different localities. On the other hand, their counterpart in the temperate waters, *Xylophaga* spp., prefers greater depths, with some species enjoying a wide range of distribution from shallow waters to deep sea. The sphaeromatids are highly destructive in the intertidal area of the wooden structures (Figure 16),

FIGURE 16: A JETTY PILE OF 'MATTI'(TERMINALIA SP.) HEAVILY DAMAGED BY PILL-BUGS AND PIDDOCKS AT THE INTERTIDAL AREA (LOCALITY : KARWAR, INDIA).

whereas the limnoriids prefer to attack in large numbers in wood installed at deeper levels (Clapp & Kenk 1963; Santhakumaran 1984, 1994). An exception to this general pattern of vertical distribution of shipworms is reported in the case of *Psiloteredo megotara* (Hanley) in Norwegian waters, where the intensity of deterioration caused by this species is found to decrease with increasing depth (Santakumaran 1984). The availability of so many different species to bring about rapid destruction of timber structures right from the littoral waters is not only ecologically significant but also economically important to all countries having an expanding shipping and fishing industry or having unprotected waterfront structures made of wood that can be destroyed in less than nine months (Turner 1947). From the archaeological point of view, the obvious observation is that even the archaeological materials lying in great depths on the sea-bed are prone to the deteriorating process of these organisms.

Shipworms are known to infest test blocks exposed up to a maximum depth of 200 metres (Tipper, 1968, as quoted by Turner, 1972b), though, in a rare instance, *L. pedicellatus* and *T. matocotana* have been extracted from wood excavated from 560 – 580 metres depth (Pailleret et al., 2007). Intense infestation, however, is noticed only in the littoral waters, probably to a depth of 30 to 40 metres. Members of the pholadid subfamily, Martesiinae, are also known to attack fresh wood mainly in the near-shore area up to a depth of 15 to 25 metres in tropical waters. As against this, Xylophagainae are generally benthic deep-sea borers, generally confined to depths greater than 150 metres (except in higher latitudes) and their known depth range is from the low-tide level at Mill-Port (Scotland) and at Trondheim (Western Norway) to 7290 metres in the Banda Trench, off Ceram (Knudsen 1961; Santhakumaran 1984; Turner 1972b; Voight 2007). Among the crustacean borers, sphaeromatids, as mentioned earlier, are mainly intertidal, infestation sometimes extending to a few metres down in shallow coastal areas (Pillai 1961). On the other hand, species of limnoriids have been recorded from intertidal areas onwards up to a depth of 1100 metres (Cookson 1991; Santhakumaran 1984).

Some of the deep water borer species have also been occasionally reported from drift-wood, and such very rare instances are the result of infested wood pieces being lifted off the bottom to the surface during storm and tossed to the littoral sea by waves and currents (Turner 1972b). For a long time, it was believed that wood pieces with borers, dredged out from deep waters, are those which had larval infestation while floating on the sea surface before getting sufficiently water-logged to sink to the bottom. Similarly, endemic species described from wood materials dredged occasionally, are limited in their range of distribution due to poor dispersal efficiency of larvae (Knudsen 1961) and also due to the occurrence of plant debris as discontinuously distributed 'islands' (Turner 1972 a). Nevertheless, fresh panels, introduced at 3000 metres in mooring line of a buoy, set out by Woods Hole Oceanographic Institution at 39° 30' N and 69o 40' W (on the Gay Head, Martha's Vineyard-Bermuda transect) were found to have been heavily infested by *Xylophaga* (as much as 100 individuals per square centimetre), which were too small at the time of retrieval to permit positive identification. This is the first instance of borer attack on 'new' wood installed at such great depth (Turner 1972a). Subsequent panel tests in deep waters not only brought out the existence of two new genera and four new species endemic to deep sea, but also indicated their intensity of attack resulting in rapid destruction of freshly arrived wood (Turner 1972a, 1972b), pointing to the potential damage they can cause to wooden wrecks lying even at such great depths. Recently, Voight (2007, 2009) also added eight more species of deep-water *Xylophaga* from wooden test blocks and materials from 1540 to 3222 metres in the Pacific Ocean, suggesting that these opportunistic borers at great depths have the unique ability to spread to any given area in the ocean floor and make use of whatever woody samples are available at their disposal.

Increased tempo in research on deep sea wood-borers since 1961, using test panels, has thus brought to light several new endemic species and has disproved the earlier belief that borers found in wood dredged from deep sea are those which infest floating wood, which sink to the bottom after water logging. These panel tests have shown that (1) intensity of attack increases with time; (2) attack is most severe very close to the mud-level; and (3) scarcity of wood and its patchy distribution in great depths has resulted in isolation and speciation. This also shows the magnitude of their threat to ship-wrecks lying at any depths in the oceans.

Salinity – briefly mentioned earlier – is an important factor to take in consideration, when studying the distribution of shipworms, as its seasonal variations can impact on the species survival and reproduction (Blum 1922), though they are generally tolerant of very low salinity level. At the same time, they will not survive when exposed to fresh water for long periods. *T.navalis* has been said to be intolerant of brackish waters, and some outbreaks in The Netherlands in the eighteenth and nineteenth centuries were explained due to reduced rainfalls that resulted in raising the salinity content in the water (Calman 1919). However, its presence seem to be more or less established nowadays and recorded in these locations again in the early 2000 by Palma, during the MoSS Project, who noticed presence of wood-borers in the brackish waters of the Baltic Sea (Palma 2005c).

Chapter 5. Recent Findings

Similar research methods were, in the recent past, employed on the wreck site of Henry VIII's warship *Mary Rose*. Raised in 1982, the remains of the ship already showed deterioration by wood-borer action, where timber had been exposed without sedimentation. The chronology of the deteriorating events on the *Mary Rose* is not yet known, but certainly after the collapse of the port side of the ship, timbers were left exposed leaving them vulnerable to the environmental threats. In 1981, Mallison and Collins from the Department of Oceanography at Southampton University wrote a report on the biological survey at the *Mary Rose* excavation site, stating that the presence of *T. navalis* was suspected 'due to numerous calcareous tubes…found in the decayed upper ends of the ship's frames…' and furthermore 'The upper, near surface timbers are in very poor condition. Even below the current sea-bed level, wood was vulnerable to attack prior to the excavation due to movement of bottom sediments and penetration by organisms through the uncovered surface. These upper timbers appear spongy, and surface growth is limited because the outside layers are constantly falling away' (Mallinson and Collins 1981). It is to be supposed that Mallinson and Collins are referring here to *Teredo sp.* rather than naming the specific species. Also, although the presence of gribble was considered the main agent of attack, perhaps it was overlooked that gribbles do not line the tunnels, in which they live, with calcareous substance.

In 2003, with funding provided by the Ministry of Defence, the investigation of the site started again and continued till August 2004, when on some of the newly retrieved timbers, which were analysed under the microscope, signs of possible attack were discernible. During the investigation, the pallets and shells of different shipworms were collected and identified. It soon became evident that they did not belong to *T. navalis*. X-ray investigation supported this and it soon became evident that a different species of shipworm, *L. pedicellatus,* was present (Palma 2004). Larvae were recorded in the brood pouch of anadult, causing alarm as *Lyrodus* has a more active reproducing pattern and it is more tolerant to low temperatures than *T. navalis*. *L. pedicellatus* was found in isolated locations along the South coast of England in the Fifties and Sixties in Shoreham and Newton Ferrers (UK) (Hall and Saunders, 1967) and in the Seventies (Board and Feaver 1973) but without indicating any specific area and, of course, throughout the warmer seas of the world (Turner 1966). Presence of *L. pedicellatus* was also recorded by Board in 1962 in the upper river Hamble near Southampton (Coughlan 1977), but no evidence of this species was recorded in this area subsequently in the Seventies.

Whilst *T. navalis* reproduces during the summer months (Turner 1966), *Lyrodus* generally reproduces from October to May, with a peak from November to January (Roch 1957). Being viviparous, *Lyrodus* brooders spawn ready to settle larvae of pedi-velliger stage, which, within a few hours after release, colonise the same wood or those nearby in the area, whereas *T. navalis* larvae are free swimmers for a short period during which – with their power of locomotion - they are easily distributed by ocean currents (Turner 1947). Thus, in their planktonic stage, larvae of *T. navalis* may be carried far and wide for up to 3 weeks away from their birth location and with the opportunity of infesting other uninfected wood (Calman 1919). Edmondson refers to a research findings, published by Miami University, which conclude that larvae of *Teredo* can enter wood during the first five days after being released in the water and, if this is applicable to Teredinidae in general, the radial spread of larvae into the ocean from shore may equal or exceed a half mile each day at least (Edmondson 1962). The obvious consequence of this process is that, whilst *T. navalis* can reach other sites, where more wood is available, *L. pedicellatus* repeatedly colonises the same timber, until the entirewood substratum is exhausted. If conditions are favourable, several generations can spawn each year, so that each individual can produce huge numbers of larvae ranging from a few thousand to several million. Larger numbers of larvae obviously leads to a significantly greater amount of degradation of the surrounding wood deposits.

Temperature is one of the key factors influencing the spawning and, whereas species in warmer waters may continue throughout the year, species in cold waters may have a reduced activity (Lopez-Anido et al. 2004), which sometimes might cease altogether (Hurley 1959). According to Lopez-Anido's research on deterioration of pine in Maine, the presence of shipworms in specific locations 'and their aggressiveness contradicts the general pre-conception that shipworms are not active in cold waters'. This has already been established earlier in a series of publications by Santhakumaran (1984) and Santhakumaran and Sneli (1984).

The widespread presence of *Lyrodus* in British waters is potentially considered an indication of environmental change. The increase in the air/water temperatures may create the conditions good enough for non-indigenous species to settle and reproduce in areas like the south of England.

Floating wooden structures - vessels, driftwood - or even bilge water play an important role in the geographical distribution of these species of wood-borers. Cross contamination is quite likely to take place, also considering the amount of vessels participating in maritime events throughout Northern Europe – for example *The International Festival of the Sea* in Portsmouth (UK).

Environment

During its sailing days and also in the unfortunate event of a wreck, the wood structure of a vessel undergoes continuous changes not only from the assemblage and distribution point of view, but from the structural point of view as well. After the sinking, the unsecured and lighter object will be inevitably floated away and dispersed in the locality, while the heavier and bulkier items will be left at the mercy of the environment surrounding the site of sinking (Muckelroy 1978).In fact, not just shipwrecks, but all underwater wood and other organic structures, such as piers, wharfs, harbours, groynes, lock- gates, lying on the seabed or in a tidal environment, are subject to the action of the dynamic environmental factors, such as currents and tides, sedimentation including all types of physical, chemical and biological damage.

When archaeological material is buried under the sediment, this protective sediment layer can be disturbed by crustaceans and fish, which create large burrows (Ferrari & Adams 1990). This will affect the protective action afforded by the sediment to the underlying materials. In the event of attack by deteriorating organisms preceded silting, this action has severe consequences in the case of an anoxic environment, as these newly created burrows are fundamental for degrading organisms and fungi to have that provision of oxygen for their survival and, therefore, for continuing their attack.

The sediment characteristics have a major influence in terms of burrowing depending on the quality – if muddy or coarse – as well as the depth of the sediment layer deposited on top of the wrecks. The top 300mm of sediment in muddy environment is where most biogenic alteration would appear to happen (Ferrari & Adams 1990) and, therefore, it is a well-researched fact that archaeological specimens are more stable with limited oxygen provision below 0.5 m (Gregory 2004; Palma 2005c).Instead, if the wreckage is protruding off the seabed and exposed to aerobic conditions, there are a number of actions – human impact and natural impact - that can alter the new status of the artefacts. Although a detailed discussion on the human impact is beyond the scope of this article, a casual mention is still proper, as it is an issue related to the natural impact. The former can be intentional or accidental and may include a variety of situations, ranging from possible salvage operations subsequent to sinking, exploitation or development of sea-bed by dredging, fishing, recreational (sport divers) and cultural (research) activities. Sometimes, a beneficial consequence of human impact is the very discovery of artefacts, if not the shipwreck itself. However, more often than not, these human interventions act as a catalyst too for increased rate of deterioration of the materials. The alteration of the *equilibrium,* thus reached, might transform the site into a favourable environment for the degrading organisms to survive and reproduce, by way of providing the much needed accessibility to substratum, food and oxygen requirements.

The natural component involves the interaction of different environmental factors, which determine the composition of the benthic fauna (Rasmussen 1973). Physical, biological and chemical components influence the site stability or preservation potential of an area of archaeological interest and are the factors, which influence the presence and activity of certain marine species depending on such changes in the environment. Physical processes, caused by hydraulic action of waves and tides, are responsible for structural damage, erosion and fluctuations in sediment deposition, especially in shallow sites (O'Shea 2002). All these have a direct link with the preservation potential of the artefacts as well as with the presence of diverse organisms. Chemical processes, which include a series of parameters that are responsible for the nature of the surrounding seawater, govern the biological processes as well. For these reasons, it is important to record as much as possible the environmental parameters of the water, when researching the extent of wood-borer activity.

From what is discerned from the depth range of marine wood-borers, there is a perceptible link between their occurrence and the threat caused by them to wrecks lying in greater depths. Expectedly, the presence of wood materials and length of free-swimming larval stages decide the distribution of borers in littoral and deep waters. While wooden structures of various kinds are aplenty along the coastal area, these substrata are a rare commodity in deep sea and their occurrence is patchy. Sources for remnants of wood material in deep sea are from waterlogged floating pieces and logs, which eventually sink to the bottom, and also from shipwrecks. However, the discovery of large number of new endemic species of xylophagainids from great depths with reports on their intense settlement and destruction of wood is further proof for the immense threat paused by borers to archaeological materials lying in great depths too.

The Swash Channel Wreck

In March 2006, the execution of the project was passed to Bournemouth University (U.K.), where a where a monitoring phase was undertaken. The aim was to establish the exact nature of deterioration and obtain information for long-term management. Shipworm attack was evidenced by calcareous linings on the hull timbers, some of which measured up to 50cm (Palma & Parham 2006) (Figure 17).

FIGURE 17 HULL TIMBERS OF SWASH CHANNEL WRECK (U.K.)
DESTROYED BY MARINE BORERS.

Once again, the MoSS experience has been applied to this very dynamic site. Sacrificial wood samples were deployed at the end of May 2006, and collected at regular intervals. Furthermore, regular monthly dives were carried out to record the changes in the environment as well as sediment movements. On every dive, the site appeared dramatically changed and samples showed constant and increasing signs of bio-fouling activity. After 3 months of underwater deployment, the panels showed initial slight attack (no more than 15%) from shipworm and moderate attack (up to 10%), and in some case, severe (almost complete surface coverage) from crustaceans like *Limnoria*.

In mid-November, when the second batch of samples was collected after 6 months exposure, the condition had severely worsened. The superficial layer had been extensively degraded by the crustaceans *Limnoria* and *Chelura*, which shared the same unlined tunnels, reaching a depth inside the wood of 0.6cm. Number of shipworms had increased and they had also considerably grown in size, and their attack had become moderate (no more than 25%). Tunnels are found as deep as 2cm inside the wood structure, as reported by Oliver (1961). As previously mentioned, examination under the microscope and x-raying were part of the investigative process. X-rays, however, do not show the impressive number of overlapping young adults just few millimetres far from each other, as their calcareous parts are still not well developed and not big enough to be visible. Yet, it must be borne in mind that these young ones will, in due course, increase their size very quickly and consequently, will enlarge their tunnels inside the wood structure resulting in remarkable destruction.

As *Limnoria* degrades the superficial layer of the wood structure, the shipworm larvae find it easier to penetrate, where the wood has already been partially softened or conditioned. Although the reproductive cycle of *Teredo* is during summer months, it was noted that, even in the middle of November, live adults and larvae have been very active, involving themselves in some sort of competition for colonising the same available substratum. This might point to an obvious acquired evolutionary tolerance to environmental changes, whereby species, which were not supposed to be alive in the winter months, could be now thriving and reproducing. Examination of the pallets collected confirmed once again that they belonged to *L. pedicellatus*, and not to *T. navalis*. It is evident that we are witnessing an increased invasive activity of this species along the British coast and that its further spreading would render the exposed wooden structures or wreck in the area vulnerable to fast deterioration from biological activity.

Another observation, possibly linked to this phenomenon, is the change in water temperature. Looking at the sea temperature during the last 20 years or more, there has been a relatively slow but progressive increase in the temperature (Jones et al. 1999).First-hand experience during the November dives showed that the water temperature was 14°C, which of course, favours the survival and reproduction of these borers, which require warmer waters, compared to the more common native species.

Chapter 6. Conclusions

Submerged wood and, more particularly, surviving archaeological wood deposits can be seen to be at an even increased risk of deterioration at the present time, than at any recorded time in the past. Because, native wood-boring species are being joined and 'reinforced' by more voracious species, encroaching even further north, aided by the significant environmental changes manifested by the rising sea water temperature, as a result of general global warming. It is therefore imperative that appropriate methods for wood conservation on the sea-bed are evolved, and through continued monitoring, effective remedial practices are perfected, if our Underwater Cultural Heritage is to be saved for analysis and for future generations.

The presence of susceptible and infested timber is fundamental to sustain the destructive activity of wood-borers by providing a perennial supply of larvae for continuous infestation. Wooden hulks, driftwood and infested timber are a constant and potential threat for infestation (Hall and Saunders 1967). As demonstrated in some control method applied in the Sixties, the removal of all wood debris from Southampton water resulted in *Teredo* almost disappearing from the area.

The presence of *T. navalis* has been recorded in southern English coastal waters for long time, whereas the presence of *Lyrodus* in British waters can potentially be seen as a sign of environmental change due to the global warming.

Studies indicate that shipwrecks which are not buried under at least 0.5 m of sediment are more likely to be attacked and degraded by these woodborers. Current research on several archaeological sites (for example the Swash Channel Wreck site in Dorset) is showing the presence of all, or at least some of the species mentioned above on most sites, which makes the stability and preservation of archaeological wood very vulnerable.

Many authors in the past estimated that life expectancy of exposed archaeological timbers is about 10 years and will not exceed 20 (Skowronek 1984). Judging by the preliminary results of recent research on wood-borers activity on shipwrecks in relation to changes in the environmental parameter (such as increase in temperature for example), this statement seems to be an overestimation of the resident time of woody materials in the sea bed and is not applicable as time passes. Firstly, unless the wood is inaccessible to borers due to various reasons including sedimentation, it is susceptible to continuous deterioration within months. Secondly, the appearance of new invasive borer species on the scene, influenced by the slow but dramatic changes in environmental conditions over the years, especially the increase in temperature, has effectively accelerated the vulnerability to bio-deterioration of underwater cultural heritage today. These 'Silent Saboteurs' are now capable of rapidly destroying invaluable and irreplaceable sites, due to aggressive colonisation of newly exposed sites (Santhakumaran 1988).

Despite all innovative techniques and unceasing efforts made by man, it has not been possible to contain the marine borer hazard and timber exposed in the sea rarely escapes bio-deterioration. Only such wrecks as those buried in mud or sand or beneath a huge cargo will remain unaffected and preserved for posterity. Otherwise, borers will start feasting on the ship almost from the day it sank, and the survival of sunken ships even as wrecks depends on the mercy of wood-destroying marine pests, which may turn these 'port-holes to history into meaningless junks'. An insight into what would have happened to ancient wooden ships can be obtained from the condition of an 'Arab Dhow', run aground for about 18 months in the Mandovi estuary (Goa, west coast of India). During this time, this ship, constructed entirely of teak (*Tectona grandis* Linnaeus), was so much infested by countless shipworms and piddocks (Figure 18), that its damaged keel started crumbling due to its own weight on account of impairment of strength. When it was finally floated for possible repair work, water gushed in through several borer holes. (Figure 19)

FIGURE 18: AN ARAB DHOW, CONSTRUCTED OF TEAK (TECTONA GRANDIS), DURING LOW-TIDE IN PANAJI BEACH (GOA)

FIGURE 19: CLOSE –UP OF KEEL AND SIDE PLANKS OF ARAB DHOW SHOWING INTENSE INFESTATION OF MARINE WOOD-BORERS.

The ship could be kept afloat only after plugging the holes with cement plaster and, subsequently, replacing all the damaged planks after extensive repair work. However, the idea of salvaging the boat was eventually discarded, as it was found beyond any repair and renovation.

In the light of such catastrophic capabilities of these animal pirates of the sea in destroying timber with remarkable rapidity, details on their hazards in early navigation, furnished above, are to be considered only a fraction of events that might have been recorded in history. Probably, there could be numerous, but unrecorded, tragic instances, where ships might have fallen prey to the destructive activities of marine wood-borers. Whatever little information, available from accounts on ancient voyages, is sufficient to illustrate the harrowing experience of those early mariners due to the presence of their unseen animal foes in the sea, and also to highlight the glaring inadequacies and utter ineffectiveness of all methods employed by man to combat the menace of these silent animal saboteurs of the sea for sufficiently long periods.

Marine wood-borers, being instrumental in causing ship-wrecks, might have enriched the sea bottom, over the years, with materials for further archaeological studies. Even then, as mentioned earlier, the same organisms will continue their work on sunken ships as well, and will convert them to irretrievably useless junks. The ruthless attack of borers on timber in high intensity and the unbelievable quickness of destruction, combined with the futility of ancient protective devices employed, therefore, make it imperative that marine archaeological investigations at known shipwreck sites should be augmented to unearth invaluable historical data and artefacts, before they are lost to satisfy the insatiable appetite of these menacing marine marauders of the sea.

The deterioration process by wood-borer activity is irreversible and a constant supply of these organisms endangers all unprotected submerged wood, including archaeological deposits. We are still facing the same problems our ancestors encountered when trying to protect or clean their wooden hulls from these organisms, as an effective solution to prevent their attack has not been evolved.

Field observations indicate that the degradation by marine borers needs to be investigated and constantly monitored. The aim is to understand the potential threat and the nature of degradation. When this step is achieved, a mitigation policy can be applied to protect and preserve *in situ* the submerged cultural heritage from the deleterious effect of 'global worming'.

APPENDIX I

SYSTEMATIC TREATMENT OF MARINE WOOD-BORERS

Phylum: Mollusca
Class: Bivalvia (Pelecypoda)
Order: Myoida (Pholadacea)
Sub-order: Pholadina
Family: Teredinidae
Sub-family: Bankiinae

Pallets segmented in structure, paddle-like, segments fused and indistinct and having lateral awns (particularly in young ones) or long with fused but distinct segments or with distinct cones with lateral awns.

Genera included are :*Bankia* Gray, *Nausitora* Wright, *Nototeredo* Bartsch and *Spathoteredo* Moll.

Sub-family: Teredininae

Pallets variable but non-segmental in structure (sometimes with a pronounced periostracal cap) or paddle-shaped and non-segmental with a thumb nail-like depression at the distal end of the blade.

Genera included are :*Bactronophorus* Tapparone-Canefri, *Dicyathifer* Iredale, *Lyrodus* Gould, *Neoteredo* Bartsch, *Psiloteredo* Bartsch, *Teredo* Linnaeus, *Teredora* Bartsch, *Teredothyra* Bartsch, *and Uperotus* Guettard.

Family: Pholadidae
Sub-family: Martesiinae

Shells lack protoplax; number of other accessory plates variable (in addition to metaplax and hypoplax, siphonoplax also being present in some species); apophyses present; umbonal-ventral sulcus well developed dividing the shell valves into two distinct areas, the anterior portion with toothed ridges in wood-borers; shells in young stages beaked and gaping anteriorly, but closed by a callum in the adult. Foot atrophies in the adult. Entire body can be retracted within the shell.

Genera of wood-borers included: *Martesia* Sowerby and *Lignopholas* Turner.

Sub-family: Pholadinae

Callum absent, number of accessory plates variable, but hypoplax and siphonoplax always lacking; apophyses present; umbonal ventral sulcus absent; well-developed foot retained in adult stage as well; animal cannot completely retract within the shell; very rarely found in the wood.

Genera of wood-borers :*Barnea* and *Pholas* Linnaeus

Sub-family: Xylophagainae

Callum and apophyses absent; shell beaked, with beak at nearly right angles giving the shell a teredo-like appearance; umbonal-ventral sulcus and ridge well developed; anterior portion of shell with finely denticulate ridges and posterior portion sculptured with growth-lines; only accessory plate present is a divided mesoplax (except in Xylopholas and Xyloredo, where a pair of siphonal plates is also present); animal capable of complete retraction within the shell (except in Xylopholas and Xyloredo, where the body is elongate and teredinid-like); foot not atrophying in the adult stage.

Genera included are: *Metaxylophaga* Taki and Habe, *Neoxylophaga* Taki and Habe, *Xylophaga* Turton , *Xylopholas* Turner and *Xyloredo* Turner. Turner (1955) placed *Metaxylophaga* and *Neoxylophaga* as synonyms of *Xylophaga*.

(See Taki and Habe, 1950; Turner, 1955, 1966, 1972a, 1972b; Turner and Santhakumaran, 1989, for description of molluscan genera).

Phylum: Arthropoda
Class: Crustacea
Order: Isopoda
Sub-order: Flabellifera
Family:Sphaeromatidae
Sub-family: Sphaeromatinae

Includes wood-boring isopods with pleopods 4 and 5 having transverse respiratory folds in the endopodites only, exopodites being thin and membranous; adult measuring about 7 to 14 mm; endopod of uropod is immovably fixed to the protopod; maxilliped lacks epipodite; first five pleonal segments partly fused with incomplete suture lines in the margin of the pleon indicating original segmentation; sixth pleonal segment fused with telson to form pleotelson; posterior part of the body with characteristically placed tubercles with or without setae.

Wood-boring genus: *Sphaeroma* Latreille.

Family: Limnoriidae

Wood-boring and algal (hold-fast) boring isopods, adult measuring about 2 to 7 mm; exopod of uropod much shorter than endopod, endopod without apical claw; first antenna with 4 or fewer flagellar articles; maxilliped with epipodite; first peraeopod with secondary unguis bifid or undivided and sometimes with accessory spinules; pleon segments separate; pleotelson circular or oval and 5th segment and pleotelson with lateral crests or latter with small tubercles.

Wood-boring genera: *Limnoria* Leach and *Paralimnoria* Menzies.

(See Harrison and Holdich 1984; Pillai 1961; Cookson 1991; for description of isopod crustacean genera).

Order: Amphipoda
Sub-Order: Corophiidea
Super-Family: Cheluroidea
Family: Cheluridae

Wood-boring amphipods, often found in association with *Limnoria*. In this, the last three pleon segments are large, immovably fused and marked ventrally by sutures, the third segment being very large. Uropods I,II,III are dissimilar to each other in shape and size. Accessory flagellum present on antenna I; flagellum of antenna II consisting mainly of a single large article in the adult.

(See Barnard 1959; Ortiz 1976, for description of chelurid species)

APPENDIX I

CHARACTERS OF TAXONOMIC VALUE FOR IDENTIFICATION OF MARINE WOOD-BORERS

Shipworms (Teredinidae): For a long time, classification of teredinid species (shipworms) was based entirely on shells and pallets, although variations in specimens of the same species were confusing. It was Turner (1966) who brought out the importance of the anatomy of shipworms in the systematics of this group, particularly in generic classification. Characters of systematics value for species identification are nature of the shell valves, tubes (internal lining of the borrows which sometimes gets thickened as a tube particularly at the posterior end), pallets (a pair of calcareous organ situated at the posterior end of the animal which is used to plug the entry hole during adverse conditions or when the borer is disturbed) and siphons. Of these, the morphological variations exhibited by the pallets are remarkable and almost all the species can be identified from their pallets (Turner 1966, 1971). Other characters are of limited help, but when considered with more important characters, may prove useful in separating closely related species. It will be useful to examine a series of pallets of the same species, as fresh as possible, and some characters are better discernible under transmitted light.

Piddocks (Pholadidae): The piddocks or members of the family Pholadidae are classified based on the shape of shell valves, nature and arrangement of accessory plates (protoplax, mesoplax, metaplax, hypoplax, siphonoplax and siphonal plates), presence or absence of callum in the adult stage, presence or absence of apophysis, and on the morphology of the siphons. In some members (Sub-Family Martesiinae), the young and adult are different morphologically, the former having an anteriorly beaked and widely gaping shell and the latter having this gape closed by a calcareous deposit, the callum (Turner 1971; Turner and Santhakumaran 1989). The nature of the chitinous lamellae on the posterior slope of the shell, when present, also helps in species separation.

Nomenclature of parts in members of Teredinidae and Pholadidae is shown in Figure 20 – A to G.

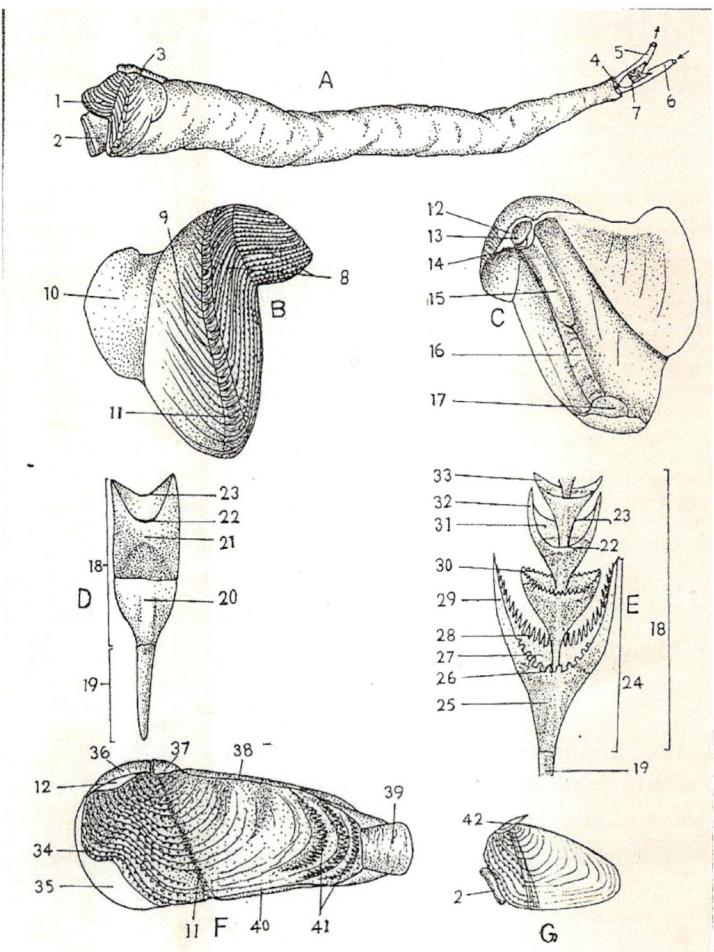

FIGURE 20: HYPOTHETICAL, COMPOSITE DRAWING OF A TEREDINID AND PHOLAD GIVING NOMENCLATURE OF PARTS. A : ENTIRE SHIPWORM; B : EXTERNAL VIEW & C : INTERNAL VIEW OF RIGHT VALVE OF SHIPWORM; D: PALLET OF TEREDO; E: HYPOTHETICAL PALLET OF BANKIA; E : HYPOTHETICAL PHOLAD; G YOUNG PHOLAD. 1 : SHELL; 2 : FOOT; 3 : CEPHALIC HOOD; 4 : MANTLE COLLAR; 5 : EXCURRENT SIPHON; 6 : INCURRENT SIPHON; 7 : PALLET; 8 : ANTERIOR SLOPE; 9 : DISC; 10 : POSTERIOR SLOPE; 11 : UMBONAL-VENTRAL SULCUS; 12 : UMBONAL REFLECTION; 13 : DORSAL CONDYLE; 14 : CHONDROPHORE; 15 : APOPHYSIS; 16: UMBONAL-VENTRAL RIDGE; 17 : VENTRAL CONDYLE; 18 : BLADE; 19 : STALK; 20: CALCAREOUS BASE; 21: PERIOSTRACAL CAP; 22 : OUTER MARGIN; 23 : INNER MARGIN; 24 : CONE; 25 : CALCAREOUS PORTION; 26 : COARSE SERRATIONS; 27 : PERIOSTRACAL PORTION; 28 : COMB-LIKE SERRATIONS; 29 : SERRATED LONG AWN; 30 : FINE SERRATIONS; 31 : WEB; 32 : NON-SERRATED LONG AWN; 33 : BROAD SHORT AWN; 34 : BEAK; 35 : CALLUM; 36 : PROTOPLAX; 37 : MESOPLAX; 38 : METAPLAX; 39 : SIPHONOPLAX; 40 : HYPOPLAX; 41 : PERIOSTRACAL LAMELLAE; 42 : FALANGE.

Pill - Bugs (Sphaeromatidae): Characters of taxonomic value in species of Sphaeromatidae are the number and disposition of large tubercles on the dorsal posterior part of body, on posterior part of the telson (Figure 21) and shape of the epistome. Of these, the arrangement of the large tubercles is strikingly different and shows variations characteristic of each species (Pillai 1961).

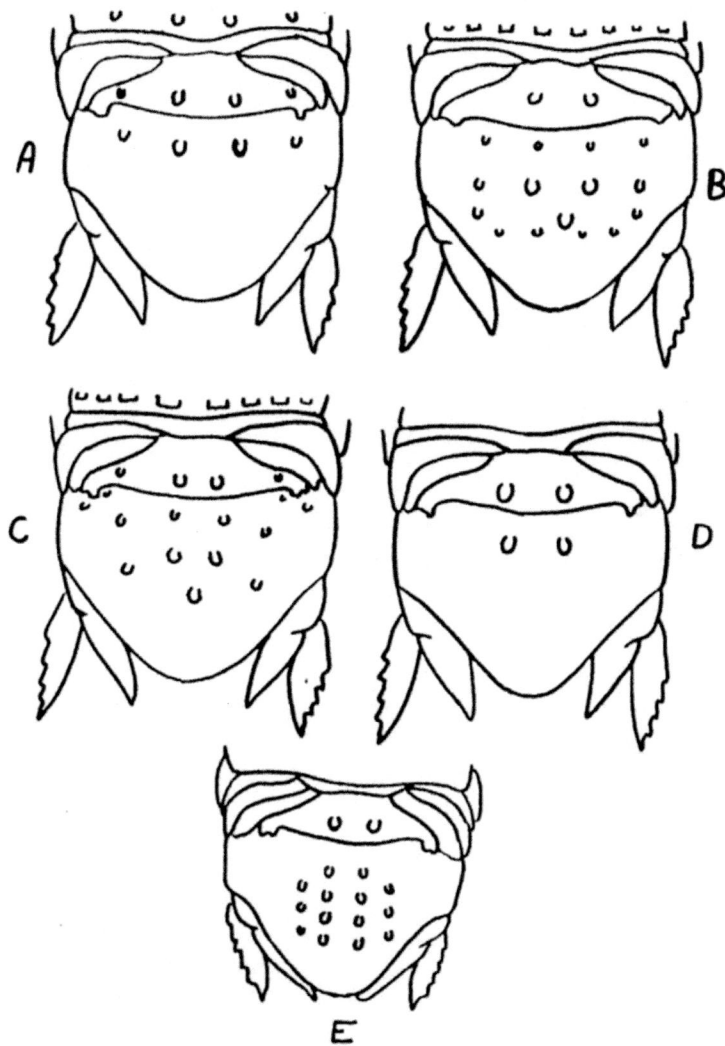

FIGURE 21: PLEOTELSON OF SPHAEROMA SPECIES SHOWING THE DISTRIBUTION OF TUBERCLES. A : S. TEREBRANS; B. *S. ANNANDALEI*; C : *S. ANNANDALEI TRAVANCORENSIS*; D: *S. TRISTE*; AND E : *S. TUBERCULATUM*.

Gribbles (Limnoriidae): In identifying species of *Limnoriidae* also, sculpturing of the dorsal side of the fifth pleon segment and the telson is the most distinguishing character. These areas of the body have grooves, ridges, tubercles and spines, which are all very characteristically arranged in different species. The mouth parts – antennae, mandibles and maxillipeds - also show variations of taxonomic value. The nature of a peculiar setae, called lacinia mobilis, found in the right mandible is a useful diagnostic feature for species identification by experienced taxonomist. Likewise, epipod of the maxilliped - also shows characteristic variations in different species. Figure 22 gives details of external morphology of *Limnoria* and of characters used in identification. (Pillai 1961; Cookson 1991).

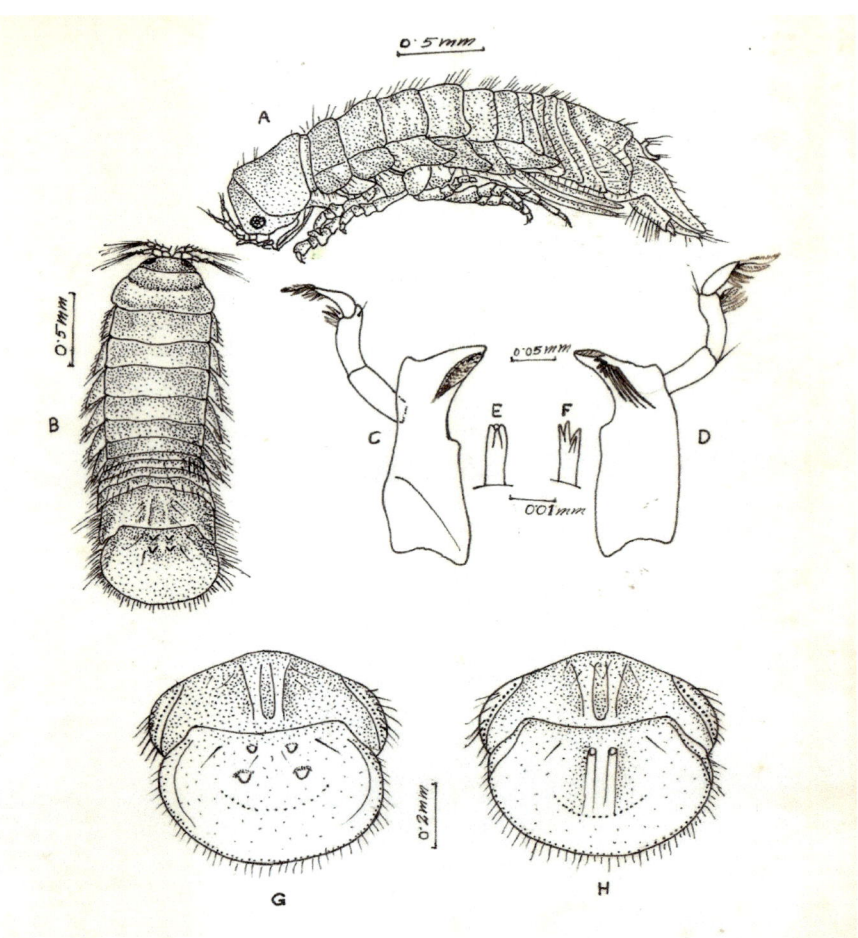

FIG. 22 : MORPHOLOGICAL CHARACTERS OF *LIMNORIA INDICA*.(A) : LATERAL VIEW AND (B) : DORSAL VIEW OF A MALE; (C) : LEFT MANDIBLE; (D) : RIGHT MANDIBLE; (E,F) : LACINIA MOBILIS OF RIGHT MANDIBLE SHOWING VARIATIONS; (G) : FIFTH PLEONAL SEGMENT AND PLEOTELSON OF MALE AND (H) OF FEMALE.

Cheluridae: Large urosome and the peculiar uropods are typical characters of this family (Barnard 1959; Ortiz 1976).

Nature of burrows produced

The burrow produced by each of the above four types of borers is also characteristic of its occupant. Shipworms bore deep into the wood, making long tunnels almost parallel to the grain (Figure 2F1), whereas burrows of pholadids are pear-shaped, superficial and nearly at right angle to the grain (Figure 2F2,3) (Some deep-water genera of the sub-family Xylophagaiinae, which possess elongate body, produce long teredinid-like tunnels). Pill-bugs produce cylindrical burrows on the wood surface at right angle to the grain (Figure 2F4). Sometimes the juveniles start working from the main parent tunnel leaving side branches (Figure 2F4b). Gribbles tunnel just below the wood surface along the wood grain. The burrows are connected to the wood surface with a series of minute ventilation holes appearing in a line (Figure 2F5). The juveniles burrow from the original parent tunnel itself, and from these secondary branches, the third generation initiate boring producing tertiary tunnels. Thus, *Limnoria* burrows have a highly branching nature and the wood surface will be spongy in texture and lace-like in appearance. Chelurids make long furrow on the wood surface often working from and enlarging *Limnoria* tunnels.

APPENDIX II

CHECK-LIST OF MARINE WOOD-BORERS

Family: Teredinidae

The following 75 species of Teredinidae (shipworms) have so far been described from all over the world (Turner 1966, 1971; MacIntosh 2012).

Sub-family Bankiinae:

1. *Bankia anechoensis* Roch, 2. *Bankia australis* (Calman) 3. *Bankia bagidaensis* Roch, 4. *Bankia barthelowi* Bartsch, 5. *Bankia bipalmulata* (Lamarck), 6. *Bankia bipennata* (Turton), 7. *Bankia brevis* (Deshayes), 8. *Bankia campanellata* Moll and Roch, 9. *Bankia carinata* (Gray), 10. *Bankia ceiba* Clench and Turner, 11. *Bankia destructa* Clench and Turner, 12.*Bankia fimbriatula* Moll and Roch, 13.*Bankia fosteri* Clench and Turner, 14.*Bankia gouldi* (Bartsch), 15.*Bankia gracilis* Moll, 16.*Bankia martensi* (Stempell), 17.*Bankia neztalia* Turner and McKoy, 18.*Bankia nordi* Moll, 19.*Bankia orcutti* Bartsch, 20.*Bankia philippinensis* Bartsch, 21.*Bankia rochi* Moll, 22.*Bankia setacea* (Tryon), 23.*Bankia zeteki* Bartsch, 24.*Nausitora dryas* Dall, 25.*Nausitora dunlopei* Wright, 26.*Nausitora fusticula* (Jeffreys), 27. *Nausitora hedleyi* Schepman, 28.*Nausitora oahuensis* (Edmondson), 29.*Nototeredo edax* (Hedley), 30.*Nototeredo knoxi* (Bartsch), 31.*Nototeredo norvagica* (Spengler), 32.*Spathoteredo obtusa* (Sivickis), 33. *Spathoteredo spatha* (Jeffreys);

Sub-family Teredininae:

34. *Bactronophorus thoracites* (Gould), 35. *Dicyathifer manni* (Wright), 36.*Lyrodus affinis* (Deshayes), 37.*Lyrodus auresleporis* Munari, 38.*Lyrodus bipartitus* (Edmondson), 39.*Lyrodus floridanus (Bartsch),* 40. *Lyrodus massa* (Lamy), 41. *Lyrodus medilobata* (Edmondson), 42.*Lyrodus pedicellatus* (Quatrefages), 43.*Lyrodus singaporeana* (Roch), 44.*Lyrodus takanoshimensis* (Roch), 45.*Lyrodus turnerae* MacIntosh, 46.*Neoteredo reynei* (Bartsch), 47.*Psiloteredo healdi* (Bartsch), 48.*Psiloteredo megotara* (Hanley), 49.*Psiloteredo senegalensis* (Blainville), 50.*Teredo aegypos* Moll, 51.*Teredo bartschi* Clapp, 52.*Teredo clappi* Bartsch, 53.*Teredo fulleri* Clapp, 54.*Teredo furcifera* von Martens, 55.*Teredo indomalaiica* Roch, 56.*Teredo johnsoni* Clapp, 57.*Teredo mindanensis* Bartsch, 58.*Teredo navalis* Linnaeus, 59.*Teredo parksi* Bartsch, 60.*Teredo poculifer* Iredale, 61.*Teredo portoricensis* Clapp, 62.*Teredo somersi* Clapp, 63.*Teredo triangularis*

Edmondson, 64.*Teredothyra dominicensis* (Bartsch) 65.*Teredothyra excavata* (Jeffreys), 66.*Teredothyra matocotana* (Bartsch), 67.*Teredothyra remiformis* (Li), 68. *Teredothyra smithi* (Bartsch), 69.*Teredothyra malleolus* (Turton), 70.*Teredora palauensis* (Edmondson), 71.*Teredora princesae* (Sivickis), 72.*Uperotus clavus* (Gmelin), 73.*Uperotus lieberkindi* (Roch), 74.*Uperotus panamensis* (Bartsch), 75.*Uperotus rehderi* (Nair).

Family: Pholadidae
Sub-Family: Martesiinae:

The following eight species of Martesiinae (Piddocks) are known to infest wooden substratum (Turner 1971; Turner and Santhakumaran 1989):

1.*Martesia (Martesia) fragilis* Verrill and Bush, 2. *Martesia (Martesia) striata* (Linnaeus), 3.*Martesia (Particoma) cuneiformis* (Say), 4.*Martesia (Particoma) nairi* Turner and Santhakumaran, 5.*Lignopholas chengi* Turner and Santhakumaran, 6.*Lignopholas clappi* Turner, 7.*Lignopholas fluminalis* (Blanford) and 8.*Lignopholas rivicola* (Sowerby).

(In addition to this, three species of the pholadid sub-family Pholadinae, namely, *Barnea birmanica* Philippi, *Barnea manilensis* (Philippi) and *Pholas chiloensis* Molina have also been reported from wood, but are of extremely rare occurrence in wooden substratum and insignificant from the wood destruction point of view).

Sub-Family: Xylophagainae:

Fifty-eight species of xylophagainids are so far reported as wood-borers from different parts of the world. (Harvey 1996; Knudsen 1961; Okutani 1975; Santhakumaran 1980, Santhakumaran and Sneli 1984; Taki and Habe 1950; Turner 1971, 1972a, 1972b, 2002; Voight 2007, 2009). Species are:

1. *Metaxylophaga suplicata* Taki and Habe, 2. *Neoxylophaga rikuzenica* (Taki and Habe), 3.*Neoxylophaga teramachi* Taki and Habe, 4.*Xylophaga abyssorum* Dall, 5.*Xylophaga africana* Knudsen, 6.*Xylophaga anselli* Harvey, 7.*Xylophaga atlantica* Richards, 8.*Xylophaga aurita* Knudsen, 9.*Xylophaga bayeri* Turner, 10.*Xylophaga bruuni* Knudsen, 11. *Xylophaga clenchi* Turner and Culliney,12. *Xylophaga concava* Knudsen, 13.*Xylophaga corona* Voight, 14.*Xylophaga depalmai* Turner, 15.*Xylophaga dorsalis* (Turton), 16.*Xylophaga duplicata* Knudsen, 17.*Xylophaga erecta* Knudsen, 18.*Xylophaga foliata* Knudsen, 19.*Xylophaga gagei* Harvey, 20.*Xylophaga galatheae* Knudsen, 21.*Xylophaga gerda* Turner, 22.*Xylophaga globosa* Sowerby, 23.*Xylophaga grevei* Knudsen, 24.*Xylophaga guineensis* Knudsen, 25.*Xylophaga hadalis* Knudsen, 26.*Xylophaga heterosiphon* Voight, 27.*Xylophaga indica* Smith, 28.*Xylophaga*

japonica Taki and Habe, 29.*Xylophaga lobata* Knudsen, 30.*Xylophaga mexicana* Dall, 31.*Xylophaga microchira* Voight, 32.*Xylophaga multichela* Voight, 33.*Xylophaga muraokai* Turner, 34.*Xylophaga nidarosiensis* Santhakumaran, 35.*Xylophaga noradi* Santhakumaran, 36.*Xylophaga obtusata* Knudsen, 37.*Xylophaga knudseni* (Okutani), 38.*Xylophaga oregona* Voight, 39. *Xylophaga pacifica* Voight, 40.*Xylophaga panamensis* Knudsen, 41.*Xylophaga praestans* Smith, 42.*Xylophaga profunda* Turner, 43.*Xylophaga ricei* Harvey, 44.*Xylophaga siebenalleri* Vioght, 45.*Xylophaga tipperi* Turner, 46.*Xylophaga tomlini* Prasad, 47.*Xylophaga tubulata* Knudsen, 48.*Xylophaga turnerae* Knudsen, 49.*Xylophaga washingtona* Bartsch, 50.*Xylophaga whoi* Turner, 51.*Xylophaga wolffi* Knudsen, 52.*Xylophaga zierenbergi* Voight, 53.*Xylopholas alternai*Turner, 54.*Xylopholas crooki* Voight, 55.*Xylopholas scrippsorum* Voight, 56.*Xyloredo ingolfa* Turner, 57.*Xyloredo naceli* Turner, 58.*Xyloredo nooi* Turner.

Family:Sphaeromatidae
Sub-family: Sphaeromatinae :

Eleven species of *Sphaeroma* have been reported as wood-borers (Harrison and Holdich 1984; Pillai 1961; Santhakumaran 2005). Species are:

1. *Sphaeroma annandalei* Stebbing, 2. *Sphaeroma annandalei travancorensis* Pillai, 3.*Sphaeroma hookeri* Leach, 4.*Sphaeroma peruvianum* Richardson, 5.*Sphaeroma quoyanum* Milne Edwards, 6.*Sphaeroma retrolaevis* Richardson, 7.*Sphaeroma serratum* Fabricius, 8.*Sphaeroma sieboldii* Dollfus, 9.*Sphaeroma terebrans* Bate, 10.*Sphaeroma triste* Heller, 11.*Sphaeroma tuberculatum* George.

(Of these, *S. retrolaevis* is very similar to *S. terebrans*, and species like *S. hookeri* and *S. serratum* are very rarely found in wood. Species so far recorded from boat hulls are *S. annandalei* and *S. terebrans*. Because of their preference to estuarine and mangrove habitats, sphaeromatids may not be important in destroying submerged wood from ship-wrecks in greater depths).

Family: Limnoriidae

Thirty-one species of wood-boring limnoriid species have so far been described (Cookson 1991; Cookson et al. 2012; Pillai 1961), they are:

1. *Limnoria andamanensis* Rao and Ganapati, 2. *Limnoria bombayensis* Pillai, 3.*Limnoria borealis* Kussakin, 4.*Limnoria carinata* Menzies and Becker, 5.*Limnoria clarkae* (Kensley and Schotte), 6.*Limnoria cristata* Cookson and Cragg, 7.*Limnoria emarginata* Kussakin and Malyutina, 8.*Limnoria faveolata* Menzies, 9.*Limnoria hicksi* Schotte, 10.*Limnoria indica* Kampf and Becker, 11.*Limnoria insulae* Menzies, 12.*Limnoria japonica* Richardson, 13.*Limnoria*

kautensis Cookson and Cragg, 14.*Limnoria lignorum* (Rathke), 15.*Limnoria magadanensis* Jesakova, 16.*Limnoria multipunctata* Menzies, 17. *Limnoria orbellum* Cookson, 18.*Limnoria platycauda* Menzies, 19.*Limnoria pfefferi* Stebbing, 20.*Limnoria quadripunctata* Holthuis, 21.*Limnoria reniculus* Schotte, 22.*Limnoria saseboensis* Menzies, 23.*Limnoria sellifera* Cookson, Cragg and Hendy, 24.*Limnoria septima* Barnard, 25.*Limnoria sexcarinata* Kuhne, 26.*Limnoria sublittorale* Menzies, 27.*Limnoria tripunctata* Menzies, 28.*Limnoria tuberculata* Sowinsky, 29.*Limnoria unicornis* Menzies, 30.*Paralimnoria andrewsi* (Calman), and 31.*Paralimnoria asterosa* Cookson.

Family: Cheluridae

Only the following four species of the amphipod wood-borer are available in literature (Barnard 1959; Ortiz 1976).

1. *Chelura terebrans* Philippi, 2.*Tropichelura insulae* (Calman), 3.*Tropichelura gomezi* Ortiz, and 4.*Nippochelura brevicauda* Shiino.

References

Adam, C., 1599. The newe nauigation and discouerie of the kingdome of Moscouia, by the northeast, in yeere 1553 ... In : *Richard Haklvyt, The principal navigations, voyages, traffiqves, and discoveries of the English nation ... London 1599*. Also new edition in 1809.

Barnard, J.L., 1959. Generic partition in the amphipod family, Cheluridae, marine wood-borers. *Pacific Naturalist*, 1 (3-4) : 3 – 12.

Bingeman, J. M., Bethell, J. P., Goodwin, P. & Mack, A. T., 2000. 'Copper and other Sheathing in the Royal Navy', *Int. J. Naut. Arch.* **29.2** : 218-229.

Bishop, F. 1913. *Panama, past and present.* New York.

Bitz, J.J., 1967. Marine Sciences and Armada. *Sea Frontiers*, 13 (5).

Blum, H.F., 1922. On effect of low salinity on *Teredo navalis*. *University of California Publications in Zoology,* 22, (4) : 349-368.

Board P.A. & Feaver, M.J., 1973. Rearing and Radiographing the Shipworm *Lyrodus pedicellatus*. In: *Marine Biology* 20: 265-268. Springer-Verlag.

Boece, H., 1527. *Scotorum historiae a prima gentis origine* ...(Paris, 1627 ?). (English translation by W. Harrison in Raphael Holinshead, the first volume of the chronicles of England, Scotlande and Irelande... Vol. 1, London, 1577).

Brown R., Bump H., Muncher D. A., 1988. An *in situ* method for determining decomposition rates of shipwrecks. *International Journal of Nautical Archaeology and Underwater Exploration* 17 (2) : 143–145.

Calman, W.T. 1919. MarineBoring Animals Injurious to Submerged Structures. By W. T. Calman. British Museum (Natural History). Economic Series, No. 10. London,.

Chellis, R.D., 1961. Deterioration and Preservation of Piles. *Piles Foundations,* McGraw-Hill, New York : 339-372.

Clapp, W.F. & Kenk, R., 1963. *Marine Borers : An Annotated Bibliography.* Office of Naval Research, Department of Navy, Washington, DC. 1136 pp.

Cnippingius, B., 1670. *Epistolae ex Ponto.* (Letters from the Black Sea).
(In: P. Ovidii Nasonis Opera Omnia, ed. by B. Cnippingius, Vol. 3 : 575-580. Leiden, 1670).

Colon, C., 1503. Letters written to the King and Queen of Spain, from Jamaica, dated July 7, 1503.

Cookson, L.J., 1991. Australian species of Limnoriidae (Crustacea : Isopoda). *Memoirs of the Museum of Victoria*, 52 (2) : 137 – 262.

Cookson, L.J., Cragg, S.M. and Hendy, I. W., 2012. Wood-boring limnoriids (Crustacea, Isopoda) including a new species from mangrove forests of the Tukang Besi Archipelago, Indonesia. *Zootaxa 0000, Mangolia Press (online edition)*, ISSN 1175 – 5334.

Coughlan, J.,1977. Marine wood borers in Southampton Water – 1951 -1975. *Laboratory note: Central Electricity Research Laboratories*, Leatherhead.

Cragg, S.M., Pitman, A.J., Henderson, S.M., 1999. Developments in the understanding of the biology of marine crustaceans and in methods of controlling them. *International Biodeterioration and Biodegradation* **43** (4): 197-205. Elsevier.

Dampier, W., 1697. *A new voyage round the world.* London, 1697. (3rd ed., in 1698).

Deslandes, A.F.B., 1768. *Essai sur la marine des anciens, et particulierement sur les vaisseaux de guerre, Paris, 1768.*

Drinkwater Bethune C.R. (ed.), 1847. *The Observations of Sir Richard Hawkins in his voyage into the South Sea, 1593.* Hakluyt Society, Vol. 1, p118-121.

Edmondson, C.H., 1955. Resistance of woods to marine borers in Hawaiian waters. *Bulletin* 217: 1 – 91. Published by Bernice P Bishop Museum, Honolulu, Hawai.

Edmondson, C. H., 1962. Teredinidae, Ocean Travelers. *Bernice P. Bishop Museum. Occas. Papers.* 23 (3): 45 – 59. Hawaii.

Eldman Abbate M. L. 1967. Primo Contributo allo studio delle alterazioni da teredini in vari legnami immersi nel Mare Ligure. In: *Contributi Scientifico-Pratici per una Migliore Conoscienza ed Utilizzazione del Legno.* Consiglio Nazionale delle Ricerche, Roma.

Ferrari, B. & Adams, J., 1990. Biogenetic modifications of marine sediments and their influence on archaeological material. *The International Journal of Nautical Archaeology and Underwater Exploration.* Technical Communication. **19**.2 :139-151.

Gianfrotta P.A. (Ed), 2000. Plinio, *Storia naturale. Libro 9, Gli animali acquatici*, Ponza, Il grande blu.

Goodell, B., 2000. Wood products: deterioration by insects and marine organisms. *Encyclopaedia of material science and technology.* Elsevier, New York.

Gregory, D., 2004. Monitoring Wooden Shipwrecks: Monitoring the Burgzaand Nood 10 and Dars Cogg using Eauxsys data logger. National Museum of Denmark. Report No. 12725-0002-01.

Hall, G.S. & Saunders M. G., 1967. Incidence of Marine Borers round Britain's Coasts. Timber Research and Development Association.

Harvey, R., 1996. Deep water Xylophagaidae (Pelecypoda : Pholadacea) from the north Atlantic with descriptions of three new species. *Journal of Conchology*, 35 (6) : 473 – 481.

Home, E. 1806. Observation on the shell of the sea worm found on the coast of Sumatra, proving it to belong to a species of *Teredo*: with an account of the anatomy of the *Teredo navalis. Proc. Zool. Soc. London*, 1912, 260-358.

Harrison, K. and Holdich, D.M., 1984. Hemibranchiate sphaeromatids (Crustacea : Isopoda) from Queensland, Australia, with a world-wide review of the genera discussed. *Zoological Journal of the Linnean Society*, 81 : 275 – 387.

Hurley, D. E., 1959. The growth of *Teredo (Bankia australis Calman)* in Ontago Harbour. *New Zealand Journal of Science*, 2 (3). New Zealand Oceanographic Institute, Contribution No. 61.

Isham, L. B. and Tierney, J. Q., 1953. Some aspects of the larval development and metamorphosis of *Teredo (Lyrodus) pedicellata* De Quatrefages. *Bull. Marine Sci.* Gulf and Caribbean, 2 (4): 574-589. University of Miami - Rosenstiel School of Marine and Atmospheric Science.

Jeffreys, J. G., 1865. British Conchology, or an Account of The Mollusca which now inhabit the British Isles and the surrounding seas. Vol. III. Marine Shells. J. Van Voorst, London.

Jones, P. D., New, M., Parker, D. E., Martin, S. & Rigor, I.G., 1999. Surface air temperature and its variations over the last 150 years. *Reviews of Geophysics* **37** : 173-199.

Knudsen, J., 1961. The bathyal and abyssal *Xylophaga.Galathea Report*, Volume 5. Danish Science Press, Copenhagen : 162 – 209.

Kuhne, H. & Becker, G., 1964. Der Holz-Flohkrebs *Chelura terebrans* Philippi (Amphipoda, Cheluridae). *Beih. Z. Angew. Zool.* No.1 : 141 – 152.

Kuenzel, J.G., 1951. Recent progress in preventing marine borer damage to hulls of wooden ships. In : *Proc. Wrightsville Beach Marine Conference, 1950*. National Research Council, Publ. 203.

Lopez-Anido, R., Michael, A., Goodell, B., and Sandford, T. (2004). "Assessment of Wood Pile Deterioration due to Marine Organisms." *J. Waterway, Port, Coastal, Ocean Eng.*, American Society of Civil Engineers. 130(2), 70–76

Major, R.H., 1847. *Select letters of Christopher Columbus ... relating to the four voyages to the New World.* (Hakluyt Society, Vol. 2: 204 – 234. London 1847). (English translation from the Spanish work of Memdez, D.: Relacion hecha por Diego Mendez, de algunos acontecimientos delultimo viage del Almirante Don Cristoba Colon, 1536).

Mallinson, J.J. and Collins, K.J.. 1981. A biological survey of the *Mary Rose* excavation site – May-December 1981. Second Report. Department of Oceanography, University of Southampton.

MacIntosh, H., 2012. *Lyrodus turnerae*, a new teredinid from eastern Australia and the Coral Sea (Bivalvia : Teredinidae). *Molluscan Research*, 32 (1) : 36 – 42.

McKee, A.1973, *King Henry VIII's Mary Rose: Its Fate and Future: The Story of One of the Most Exciting Projects in Marine Archaeology*. Souvenir Press.

Miller, R. C., 1922. Variations in the shell of *Teredo navalis* in San FranciscoBay. *Publication in Zoology*, 22 (2) : 293 – 328. University of California.

Miller, R. C., 1923. Variations in the Pallets of *Teredo navalis* in San Francisco Bay. *Publication in Zoology,* 22 (8) : 401-414. University of California.

Miller, R. C. 1924. The boring mechanism of *Teredo. Publication in Zoology,* 26, (4) : 41-80. University of California.

Moffett, T., 1634. *Insectorum sive minimorum animalium theatrum*.326 pp. Londini, 1634. (English translation in 1658).

Monson, W., 1682. A true and exact account of the wars with Spain, in the region of Q, Elizabeth (of famous memory).In : *Megalopsychy*, Part I, (5) +55, London, 1682.

Mountford, K., 2002. Fouling up: The trials and errors of protecting ships' hulls. *The Bay Journal* 12 (July-August).

Morrison,J.S., Coates,J.F., Rankov, N.B., 2000.*The Athenian Trireme.* Cambridge University Press.

Muckelroy, K., 1978. *Maritime Archaeology.* Cambridge: Cambridge University Press.

Okutani, T., 1975. Deep-sea bivalves and scaphopods collected from deeper than 2000 m in the North-western Pacific by the R/V Soyo-Maru and the R/V Kaiyo-Maru during the years 1969 – 1974. *Bull. Takai Regional Fisheries Research Laboratory*, 82 : 57 - 87.

Oliver, A.C., 1961. An Account of the biology of *Limnoria. Journal of Institute of Wood Research*, No.7-10 : 32-91.

Ortiz, M., 1976. A new wood-boring amphipod (Amphipoda Gammaridea Cheluridae) from Cuban waters. *CienciasSerie Investigaciones Marinas (Havana)*, 27 : 21 – 26.

O'Shea, K. O., 2002, *The Archaeology of Scattered Wreck-Sites: Formation Processes and Shallow Water Archaeology in Western Lake Huron. International Journal for Nautical Archaeology*, **31** (2) : 299-305.

Oviedo Y. Valdes, G.F. de, 1547. *Coronica delas Indias : la hystoria general de las Indias agora nueuamente impressa corregida y emendada.* 192 + 23 leaves (no place), 1547.

Oxley, I. 1990 .Factors Affecting the Preservation of Underwater Archaeological Sites. IJNA 19 (4) : 340–341

Oxley I. 1996. The *in-situ* preservation of underwater sites. In: *Preserving archaeological remains in situ.* Museum of London.

Oxley, I. 1998. The In-Situ Preservation of Underwater Sites. In Preserving Archaeological Remains In Situ: Proceedings of the Conference of 1st–3rd April 1996, M. Corfield, P. Hinton, T. Nixon, and M. Pollard (editors), pp. 159–173. Museum of London Archaeology Service, Bradford, England.

Pailleret, M., Takuma Haga, Patricea Petit, Catherine Prive-Gill, Nima Saedlou, Francoise Gaill and Magali Zbinden, 2007. Sunken wood from the Vanuatu Islands : identification of wood substrates and preliminary description of associated fauna. *Marine Ecology*, 28 (1) : 233 – 241.

Palma P., 2004. *Report on the Analyses conducted on some Mary Rose timbers retrieved during summer 2004.* Mary Rose Unpublished Report, November 2004.

Palma P., 2005a. *Investigation of the Mary Rose site's environment: Research on woodborer degradation. 1ˢᵗ Report.* Mary Rose Unpublished Report, March 2005.

Palma P., 2005b. *Investigation of the Mary Rose site's environment: Research on woodborer degradation. 2ⁿᵈ Report.* Mary Rose Unpublished Report, August 2005

Palma P., 2005c. Monitoring of Shipwreck Sites, *International Journal of Nautical Archaeology,* **34** (2): 232-331

Palma P. & Parham D., 2006. Swash Channel Wreck, November, Season Report written for English Heritage

Pillai, N.K., 1961. *Monograph, Wood-boring Crustacea of India.* 61 pp. Manager of Publications, Government of India Press, New Delhi.

Purchas, S., 1625. *Purchas his pilgrimes.* 4 Parts, London, 1625.

Rao, M.V., Balaji, M., Rao, K.S. and Santhakumaran, L.N., 2007. *Biodeterioration of timber and its prevention in Indian coastal waters. Third Progress Report :*

1982 – 2005. Institute of Wood Science & Technology, Bangalore – 560003, India. 198 pp.

Rasmussen, E., 1973. Systematics and Ecology of the Isefjord Marine Fauna (Denmark). *Ophelia*, 11: 315-316, August 1973. Denmark.

Riley, H.T., 1853. *The annals of Roger de Hovenden comprising the history of England and of other countries of Europe...* Vol. 2, 556 pp. London, 1853. (English translation of *Annalium pars prior and posterior. In : Rervm anglicarvm scriptores post Bedam praeipvi...,* leaves 229 – 471. Ed. Henry Savile, Londini, 1596).

Riley, H.T., 1855. *The natural history of Pliny.* Vol. 2 (Book VI X) and 3 (Book XI XVII).London, 1855. (English translation from *"Naturalis Historia"*. 356 leaves (Venetiis), 1469 of Plinius Secundes (23 – 79 AD).

Roch F., 1957. Le Teredinidi della Laguna Veneta. E dipendenza dalle condizioni idrografiche locali per quanto riguarda la loro distribuzione geografica. *Boll. Mus Civ.St Nat.* 10: 87-111. Venezia.

Santhakumaran, L.N., 1978. Response of wood-boring and fouling organisms to preservative-treated timber exposed in the Trondheimsfjord (Western Norway). In : Biodeterioration. Proc. Fourth International Symposium, Berlin, 1978. Oxley, T.A., D. Allsopp & G. Becker, Editors, Pitman Publishing Ltd, London and the Biodeterioration Society, 205 – 211.

Santhakumaran, L.N., 1980. Two new species of *Xylophaga* from Trondheimsfjorden, western Norway (Mollusca : Pelecypoda). *Sarsia*, 65 : 269 – 272.

Santhakumaran, L.N. 1984. Vertical Distribution of fouling and wood-boring organisms in Trondheimsfjorden (Western Norway). *Gunneria*, 47. Det KGL. Norske Videnskabers Selskab Museet. Trondheim.

Santhakumaran, L.N. & Sneli Jon-Arne, 1984. Studies on the marine wood-boring and fouling organisms of Western Norway. *Gunneria,* 48 : 1 – 36.

Santhakumaran, L.N. 1988. On silent saboteurs and shipwrecks. In: *Marine Archaeology of Indian Ocean Countries. Proceedings of the First Indian Conference 12-15 October 1987* (Ed) S.R. Rao: 123-126.

Santhakumaran, L.N. 1990. Marine borer attack of a wooden ship off Dwaraka coast and the significance of such damage to archaeology. *Journal of Marine Archaeology,* 1:99 – 102.

Santhakumaran, L.N.,1991. "The Saga of the Shipworms in early Navigation." *Recent advances in marine archaeology: proceedings of the Second Indian Conference on Marine Archaeology of Indian Ocean Countries: January, 1990.* Society for Marine Archaeology, National Institute of Oceanography, 1991.

Santhakumaran, L.N. 1992. The saga of the shipworms in early navigations. In: *Marine Archaeology of Indian Ocean Countries.* (Ed) S.R. Rao: 163-166.

Santhakumaran, L.N., 1994. Marine Wood-borers of India – An Annotated Bibliography. Institute of Wood Science & Technology, Bangalore-560003, India. 262 pp.

Santhakumaran, L.N., 2005. Marine Wood Borers of India – a field manual. In: V.K. Dhargalkar & X. N. Verlecar, *Survey and inventorization of coastal biodiversity (West Coast)*. National Institute of Oceanography, Dona Paula, Goa.

Santhakumaran, L.N. & Alikunhi, K.H. 1983. Natural resistance of different species of Indian timbers to marine wood-borers in Bombay waters. *Indian Forest Bulletin,* No. 272 (n.s.). Delhi. 46 pp.

Santhakumaran, L.N., Jain, J.C. and Tewari, M.C., 1984. Performance of preservative-treated timber against biodeterioration in Indian waters. *The Internat. Res. Group on Wood Preserv. (Stockholm). Document No. IRG/WP/4106.* 31 pp.

Sellius G., 1733. Historia Naturalis Teredinis seu Xylophagi Marini, tubuloconchoidis speciatim Belgici. Apud Hermannum Besseling: Trajecti ad Rhenum.

Seymour, H. D., 1855. Russia on the Black Sea and Sea of Azov. R.E. Collection at Brigham Young University. J. Murray, London.

Shino, S., 1957. The marine wood-boring crustaceans of Japan II (Sphaeromidae and Cheluridae). *Wasmann Jour. Biol.,* 15 (2) :161 – 197.

Skowronek, R.K., 1984. Archaeological Testing and Evaluation of the Legare Archorage Shipwreck Site, Biscayne National Park, Summer 1983. National Park Service, Southeast Archaeological Center, Tallahassee, Fl.

Steinmayer, A.G. Jr. and Turfa, J. M., 1996. Effects of shipworm on the performance of ancient Mediterranean warships. *The International Journal of Nautical Archaeology*, 25 (2) : 104 – 121.

Sila Tripati, Sujatha, M., Rao, R.V. and Rao, K.S., 2005. Use of timber in shipbuilding industry: Identification and analysis of timber from ship-wrecks off Goa coast, India. *Current Science (Bangalore)*, 89 (6) : 1022 – 1027.

Sila Tripati, Ramalingeswara Rao, Sashikala, S., Rao, R.V. and Vijay Khedekar, 2009. Analysis of timber and coating material on an iron anchor recovered off Aguada Bay, Goa. *Current Science (Bangalore)*, 97 (9) : 1361 – 1368.

Swiny, H.W. and Katzev, M.L., 1973. The Kyremia ship-wreck : a fourth century BC Greek merchant-ship. *Marine Archaeology*, (ed.) D.J. Blackman, Butterworths, London.

Taki, I. and Habe, T., 1950. Xylophaginidae in Japan. In : Illustrated Catalogue of Japanese Shells. Ed. Tokubei Kuroda. Vol. 1 (7) : 45 – 47. Kyoto.

Theophrastus (374?-287?BC), 1495-98. Theophrastou peri phyton historias. *Venetiis*, 1495-98. In Greek. English translation by Arthur Hort : Theophrastus enquiry into plants...(Loeb Classical Library). Vol I. Xxviii+475 pages. London, 1916.

Throckmorton, P., 1970. *Ship-wrecks and Archaeology : The unharvested sea.* Victor Gollancz Ltd.

Tipper, R., 1968. Ecological aspects of two wood-boring mollusks from the continental terrace off Oregon. Dept. Oceanogr., School of Science, Oregon State Univ. : 1 – 137 (unpublished doctoral dissertation).

Turner, R. D. 1947. Collecting Shipworms. Limnological Society of America, Special Publication No.19. The Museum of Comparative Zoology, Harvard University, Cambridge, MA.

Turner, R.D., 1955. The family Pholadidae in the western Atlantic and eastern Pacific. Part II. Martesiinae, Jouannetiinae and Xylophagainae. *Johnsonia*, 3 : 65 – 160.

Turner, R. D. 1966. A Survey and Illustrated Catalogue of *the Teredinidae (Mollusca:Bivalvia)*. The Museum of Comparative Zoology, Harvard University, Cambridge, MA.

Turner, R.D., 1971. Identification of marine wood-boring mollusks.In : *Marine wood-borers, fungi and fouling organisms of wood* (Eds. Jones, E.B.G. and Eltringham, S.K.), OECD, Paris : 17 – 64.

Turner, R.D., 1972a. *Xyloredo*, a new teredinid-like abyssal wood-borer (Mollusca, Pholadidae, Xylophagainae). *Breviora* (Mus. Comp. Zool., Cambridge, Mass.) No.397 : 1 – 19.

Turner, R.D., 1972b. A new genus and species of deep water wood-boring bivalve (Mollusca, Pholadidae, Xylophagainae). *Basteria*, 36 (2-5) : 97 – 104.

Turner, R.D., 2002. On the sub-family Xylophagainae (Family Pholadidae, Bivalvia, Mollusca). *Bull. Mus. Comp. Zool.*, 157 (4) : 223 – 308.

Turner, R.D. and Santhakumaran, L.N., 1989. The genera *Martesia* and *Lignopholas* in the Indo-Pacific (Mollusca : Bivalvia : Pholadidae). *Ophelia*, 30 (3) : 155 – 186.

Voight, J.R., 2007. Experimental deep-sea deployments reveal diverse North-east Pacific bivalves of Xylophagainae (Myoida : Pholadidae). *J. Mollus. Stud.*, 73 (4) : 377 – 391.

Voight, J.R., 2009. Diversity and reproduction of near-shore vs offshore wood-boring bivalves (Pholadidae : Xylophagainae) of the deep eastern Pacific Ocean, with three new species. *J. Mollus. Stud.*, 75 (2) : 167 – 174.

Wharton, W.J.L., 1893. *Captain Cook's journal during his first voyage round the world, made in H.M. Bark "Endeavour", 1768 – '71 : a literal transcription of the original MSS.* IVI + 400 pp. London, 1893.

Willcox, C., 1827. On the defects caused in ship's bottom by marine animals, with descriptive remarks on some of the most destructive kinds. *Pap. Naval Architect. (London)*, 1 : 150 – 157.

Yule, H., 1903. *The book of ser Marco Polo, The Venetian, concerning the Kingdoms and Marvels of the East.*(Trans.) 3rd ed., Vol. 1.

Wessex Archaeology 2005. Poole Harbour Channel Deepening and Beneficial Use Scheme. Swash Channel Wreck Designated Wreck: Mitigation Works. Ref. 61340.02.

http://www.cru.uea.ac.uk/cru/info/warming/
http://www.nba.fi/INTERNAT/MoSS
UNESCO Convention on the Protection of the Underwater Cultural Heritage, Annex Rule 1 (2nd November 2001):
http://www.unesco.org/culture/laws/underwater/html_eng/conven3.shtml